AF751341

NOVEL ASPECTS OF NANOFIBERS

Edited by **Tong Lin**

Novel Aspects of Nanofibers
http://dx.doi.org/10.5772/intechopen.69649
Edited by Tong Lin

Contributors

Ick Soo Kim, Hoik Lee, Davood Kharaghani, Yanchuan Guo, Weipeng Lu, Huijing Zhao, Huanjie Chi, Chris Wright, Chris Mortimer, Jonathan Widdowson, Emriye Perrin Akcakoca Kumbasar, Seniha Morsunbul, Simge Alır, Zhijuan Pan, Long Tian, Tao Yan, Jie Li, Avinash Baji, Yiu-Wing Mai

© The Editor(s) and the Author(s) 2018
The rights of the editor(s) and the author(s) have been asserted in accordance with the Copyright, Designs and Patents Act 1988. All rights to the book as a whole are reserved by INTECHOPEN LIMITED. The book as a whole (compilation) cannot be reproduced, distributed or used for commercial or non-commercial purposes without INTECHOPEN LIMITED's written permission. Enquiries concerning the use of the book should be directed to INTECHOPEN LIMITED rights and permissions department (permissions@intechopen.com).
Violations are liable to prosecution under the governing Copyright Law.

(cc) BY

Individual chapters of this publication are distributed under the terms of the Creative Commons Attribution 3.0 Unported License which permits commercial use, distribution and reproduction of the individual chapters, provided the original author(s) and source publication are appropriately acknowledged. If so indicated, certain images may not be included under the Creative Commons license. In such cases users will need to obtain permission from the license holder to reproduce the material. More details and guidelines concerning content reuse and adaptation can be found at http://www.intechopen.com/copyright-policy.html.

Notice

Statements and opinions expressed in the chapters are these of the individual contributors and not necessarily those of the editors or publisher. No responsibility is accepted for the accuracy of information contained in the published chapters. The publisher assumes no responsibility for any damage or injury to persons or property arising out of the use of any materials, instructions, methods or ideas contained in the book.

First published in London, United Kingdom, 2018 by IntechOpen
IntechOpen is the global imprint of INTECHOPEN LIMITED, registered in England and Wales, registration number: 11086078, The Shard, 25th floor, 32 London Bridge Street
London, SE19SG – United Kingdom
Printed in Croatia

British Library Cataloguing-in-Publication Data
A catalogue record for this book is available from the British Library

Additional hard copies can be obtained from orders@intechopen.com

Novel Aspects of Nanofibers, Edited by Tong Lin
p. cm.
Print ISBN 978-1-78923-074-1
Online ISBN 978-1-78923-075-8

We are IntechOpen, the first native scientific publisher of Open Access books

3,400+
Open access books available

109,000+
International authors and editors

115M+
Downloads

151
Countries delivered to

Our authors are among the
Top 1%
most cited scientists

12.2%
Contributors from top 500 universities

CLARIVATE ANALYTICS BOOK CITATION INDEX INDEXED

WEB OF SCIENCE™

Selection of our books indexed in the Book Citation Index in Web of Science™ Core Collection (BKCI)

Interested in publishing with us?
Contact book.department@intechopen.com

Numbers displayed above are based on latest data collected.
For more information visit www.intechopen.com

For EU product safety concerns:
IN TECH d.o.o., Prolaz Marije Krucifikse Kozulić 3, 51000 Rijeka, Croatia,
info@intechopen.com or visit our website at intechopen.com.

Meet the editor

Professor Tong Lin received his PhD degree in Physical Chemistry from the Chinese Academy of Sciences in 1998. He has been serving as a professor and personal chair at Deakin University since 2013. Professor Lin is an active researcher in the field of electrospinning, functional fibers, and polymers. He contributes to the development of needleless electrospinning for large-scale nanofiber production and novel applications of nanofibrous materials. He has published over 225 peer-referred articles in high-impact-factor journals, 14 books, 16 book chapters, and 70 other papers.

Contents

Preface

Nanofibers show a number of novel properties and unique applications. The research on nanofibers has been a hot topic in recent decades. In the previous book, entitled *Nanofibers: Production, Properties and Functional Applications* published by InTech in 2011, research progress in nanofibers was presented. This book provides additional contents about nanofibers. Special fabrications, properties, and applications about nanofibers are introduced. I am greatly appreciative of the authors for their great contributions to nanofiber discipline.

Tong Lin
Deakin University, Australia

Section 1

Nanofiber Fabrication

Chapter 1

Mechanical Force for Fabricating Nanofiber

Hoik Lee, Davood Kharaghani and Ick Soo Kim

Additional information is available at the end of the chapter

http://dx.doi.org/10.5772/intechopen.73521

Abstract

Nanofiber has attracted increasing attention owing to its wide applications such as filtration, drug delivery, wound dressing, separator, etc. A lot of fabrication methods are developed in the last few decades, electrospinning method is the most frequently utilized method for producing nanofiber. However, electrospinning features a use of electrical field to produce nanofiber, which have obviously high production cost and a big burden on the environment. And several limitations are observed such as orientation of fibers and limited options of polymer and solvents, so many researchers try to develop more facile and more effective method for making nanofiber. In this chapter, recent developed fabrication methods, handspinning, direct writing, touch and brush spinning, are discussed and the advantages of each methods are described, respectively. They utilize a simple mechanical force instead of electrical force, which delivers great benefits to producing nanofiber such as orientation of fibers along with the force direction, reduction of every cost, availability of various options for selecting polymer and solvents, and a facility to design a pattern with high precision. Those innovative and novel methods will enable us to make functional nanofibers more effective than traditional methods; consequently, they will broaden the application of nanofibers.

Keywords: electrospinning, handspinning, touch and brush, direct writing, nanofibers

1. Introduction

Over the last few decades, nanotechnology has been attracted a lot of interest due to a potential for developing new materials and devices with a wide range of applications such as medicine, electronics, biomaterials and energy production [1–3]. Especially, nanofibers have been paid much attentions in various fields due to its unique characteristics e.g. high surface-to-mass ratio, flexibility, high porosity, ability to incorporate other materials, and other various

IntechOpen

© 2018 The Author(s). Licensee IntechOpen. This chapter is distributed under the terms of the Creative Commons Attribution License (http://creativecommons.org/licenses/by/3.0), which permits unrestricted use, distribution, and reproduction in any medium, provided the original work is properly cited.

advantages, which allow opening up promising applications such as in filtration, scaffolds, wound healing, drug delivery, protective clothing, catalyst, sensors, energy harvest and storage, composite reinforcement, and many others [3–7].

Various techniques for fabricating polymeric fibers with diameters ranging from micron to nanometer scales have been developed and explored such as melt blowing [8], dry spinning [9], and wet spinning [10]; electrospinning is the most popular method for fabricating nanofiber among them [11]. The first introduction of the electrospinning was presented by Anton in 1934 [12], and it has attracted a lot of interests up to now because it has been regarded as the most effective way to produce continuous nanofibers on a large scale and adjust the fiber diameter from nanometers to micrometers. Electrospinning technique uses interactions between fluid dynamics, electrically charged surfaces and electrically charged liquids for fabricating nanofiber [13]. Typically, electrospinning apparatus comprises a high voltage power supply, a syringe needle connected to power supply, and a counter-electrode collector as shown in **Figure 1**. In electrospinning process, a strong electrical field is used to draw a polymer solution into fine filaments. When a sufficiently high electric voltage is applied to the polymer solution, the droplets that are ejected from the tip have an electrostatic repulsion force that counteracts the effect of surface tension, allowing the droplets to be stretched out to form nanofibers. The erupted and stretched filaments are travels through the air, the solvent evaporates leaving behind a polymer fibers to be collected on an electrically grounded collector.

Although electrospinning is an attractive method for fabricating nanofiber, it has been limited for fabricating nanofibers using polymer that have poor solubility in solvents and low electrical conductivity such as polyolefins (e.g., polyethylene and polypropylene) [13]. In addition, the use of electrical stretching force makes high electrical voltage necessary, indicating that it can produce high cost and excessive use of energy in production. Moreover, applying nanofiber to electrical devices requires reproducibly locating them in specific

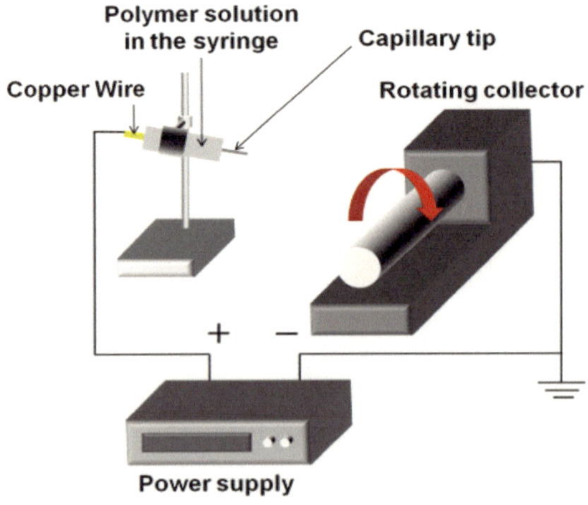

Figure 1. Schematic illustration of electrospinning apparatus.

position and orientations. However, electrospun nanofibers are difficult to apply to electrical device directly because electrospinning process typically produces fibers with random orientations without specific fabricating set-up. Thus, much effort has been devoted to developing and modifying the electrospinning method. Deitzel et al. presented a strategy that utilized electrostatic fields to control the deposition of electrospun nanofiber locating multiple field generator between tip and collector [14]. Tanase et al. reported orienting and assembling nanofibers suspended in a fluid solution using magnetic fields [15]. Zussman et al. used a disc collector instead of cylindrical collector in order to create a stronger converging electrostatic field and collect highly aligned nanofiber [16].

Apart from the electrospinning techniques, other approaches such as melt blowing, wet spinning, and dry spinning are also utilized for the fabrication of nanofibers. Melt blowing is a straightforward one step process which also one of the commonly used method for nonwovens production [17]. This method is performed by extruding a molten polymer and elongating the polymer stream coming out of the orifice by air-drag. The extrude fibers are solidified during the drawing process, and then collected on the surface of a collector. The advantages of melt blowing are high throughput rate, ease of preparing polymeric blends, suitability to the polymers having no appropriate solvent at room temperature, and unnecessary to removing the toxic solvent [18]. Even though the above advantages, melt blowing has some challenges such as the requirement of a high temperature which leads high energy cost and has possibility of polymer degradation, rapid solidification of the polymer in the orifice, and the difficulty in obtaining submicron fibers. The average diameter of melt blown nanofiber mainly depends on the throughput rate, solution viscosity, air temperature and air velocity, however, they are usually not smaller than micrometer without specially designed melt blowing setup such as using of special die with a small orifice or reducing the viscosity of the molten solution [17, 19]. On the other hands, dry spinning is used to form very thin fibers. In dry spinning process, the polymer solution dissolved in a volatile solvent is extracted through a spinneret with numerous holes (one to thousands) [20]. During the flight of the extracted nanofibers, heated air is used to evaporate the solvent so that the fibers solidify and are corrected. However, this method has been limited for selecting solvent due to safety and environmental concerns associated with solvent handling [21]. Wet spinning is the oldest method for fabricating fibers, which spinneret is submerged in a chemical bath to precipitate the fiber as it emerges [22]. A major different point of wet spinning with other spinning methods is this one which is spinning into a fluid with a much higher viscosity. The advantage of this method is that it does not require a purification process. However, wet spinning set-up is quite complex and expensive to install. In addition, this method can be applied to polymer which do not melt and dissolve only in non-volatile or thermal unstable solvents.

Herein, we introduce the recent novel developments of fabricating nanofiber apparatus; *Handspinning, Touch and Brush,* and *Direct Writing*. Those methods provide a solution that allowed us to overcome the limitations and issues in electrospinning and other spinning methods, and very simple and straightforward and versatile method that is universally applicable to any kind of polymer and solvents, it enables us to create a new type of polymeric nanofibers with exceptional properties.

2. Novel technologies for fabricating nanofibers

2.1. Handspinning

Recently, Lee et al. reported a simple and innovative method for fabricating nanofibers, called as Handspinning [23]. Handspinning was invented by mimicking commonly observed method in our daily lives such as the process of making cheese or noodles by hand-pulling, or making long and thin fiber-like structure from highly viscous liquid glue using two fingers, grabbing and subsequently pulling out the viscous material with thumb and index finger. Handspinning method relies on simple mechanical stretching force to fabricating nanofiber instead of an electrical force in electrospinning, which is a completely different mechanism for fabricating nanofiber. **Figure 2a–c** shows a schematic illustration and corresponding photographs of the nanofiber fabrication process via handspinning, and representative SEM image of a handspun nanofiber was presented in **Figure 2d**. **Figure 2e** shows that the simplest system of handspinning using only two fingers to make nanofibers.

The handspinning apparatus was designed to control processing parameters, i.e., pulling away speed (PAS, cm/s), pulling away distance (PAD, cm), and plate area (PA, cm^2), instead of voltage, tip to collector distance, feed rate which are processing parameters of electrospinning. Handspinning provides a number of options for polymers and solvents because their electrical properties are not relevant at all. Watanabe et al. reported that the diameter and surface morphologies of handspun fiber depend on solvent systems and processing conditions to control simple mechanical force [24]. They exploit a typical polyolefins, polypropylene, to investigate the utilization of handspinning. It is noticeable that polypropylene successfully fabricated in single solvent system (cyclohexane) via handspinning (**Figure 3a**), while it was not obtained

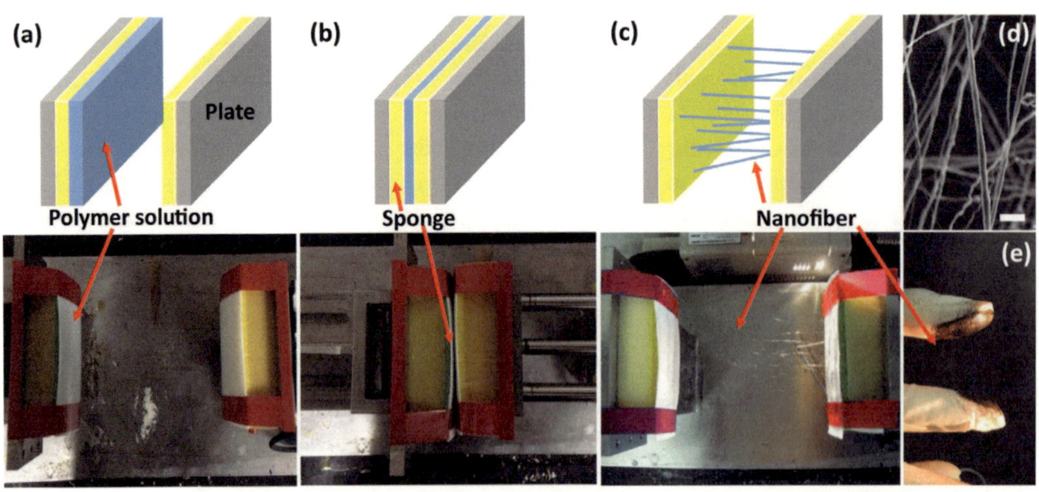

Figure 2. (a–c) Schematic illustrations and corresponding photographs of the nanofiber fabrication process via handspinning; (d) representative SEM image of a handspun nanofiber (scale bar = 10 μm); (e) photograph showing handmade nanofibers using two fingers (Reproduced with permission from Ref. [17]. Copyright 2016, Nature Publishing Group).

by electrospinning system. In addition, the morphologies of handspun and electrospun nanofibers were significantly different in same solvent system. The handspun polypropylene from the mixed solvent system shows smooth surface (**Figure 3b**), whereas the electrospun polypropylene exhibited rough surfaces with the collapsed morphology (**Figure 3c**).

Also, a significant difference in polymer chain conformation between the handspun and electrospun nanofiber was observed, and it was attributed to mechanical stretching force. The conformation of polypropylene was confirmed by FT-IR spectroscopy (**Figure 4**), which has been well-known to be a useful method to determine molecular and chain conformation. In FT-IR spectrum, it is clearly observed that the helical conformations were produced mainly when nanofiber electrospun, while hansdpun nanofiber appeared the long stands in the trans-planar conformation [24]. These results strongly suggest that the handspinning process produces more stretched fiber than the electrospun fiber, resulting in stiffer and stronger nanofiber [24].

Another advantage of handspinning is that it enables to concentrating nanofiller such as carbon nanotube (CNT), which can improve thermal and mechanical properties of nanofiber. Incorporation of a high concentration of CNTs in a polymer matrix without aggregation and localization of the CNTs is difficult because they tend to make a bundle, resulting in their being poor dispersed in the polymer matrix and the deterioration of the properties of the material [25, 26]. In addition, increasing the concentration of filler in nanofiber is limited [27]. In handspinning, the process relies only on a simple, mechanical pulling motion, hence the effect of the concentration of CNTs on the process likely is negligible. In addition, the simple mechanical stretching force in handspinning is more effective to align the CNTs in the nanofiber than the force induced by the electric field in electrospinning. **Figure 5** shows typical SEM and TEM images of electrospun and handspun PVAc/CNT nanofibers with various concentration of CNTs. It is worth to note that handspun nanofibers are aligned quite well compared to electrospun nanofibers, which typically are distributed randomly. Since the force is applied uniaxially to the polymer solution by a pull-out motion, nanofibers are fabricated along the axis in the direction of the operation, leading to well-aligned nanofibers. In addition, in TEM image, it clearly indicates that the orientation of the CNTs in the PVAc matrix was dependent on the fabrication method due to the different applied force. The pulling out from mechanical force in handspinning process affects the distribution of the filling materials in nanofiber. On

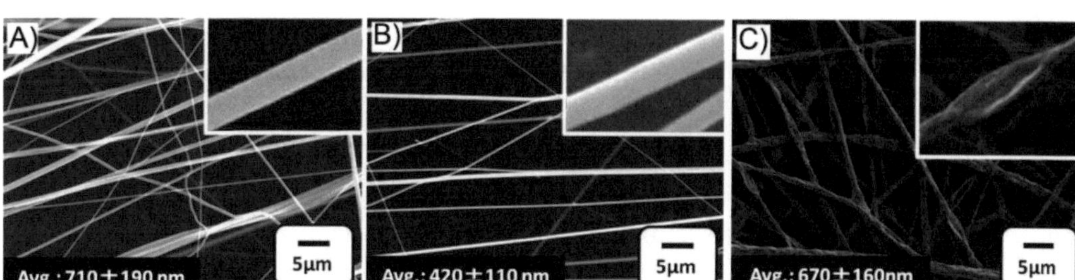

Figure 3. SEM images of handspun PP nanofibers obtained from the single solvent system (a: cyclohexane), mixed solvent system (b: cyclohexane/acetone/DMF = 80/10/10 wt%), and (c) electrospun PP nanofibers. PAD = 10 cm, TCD = 15 cm. The insets show the magnified SEM images (Reproduced with permission from Ref. [18]. Copyright 2011, Wiley-VCH Verlag GmbH & Co. KGaA, Weinheim).

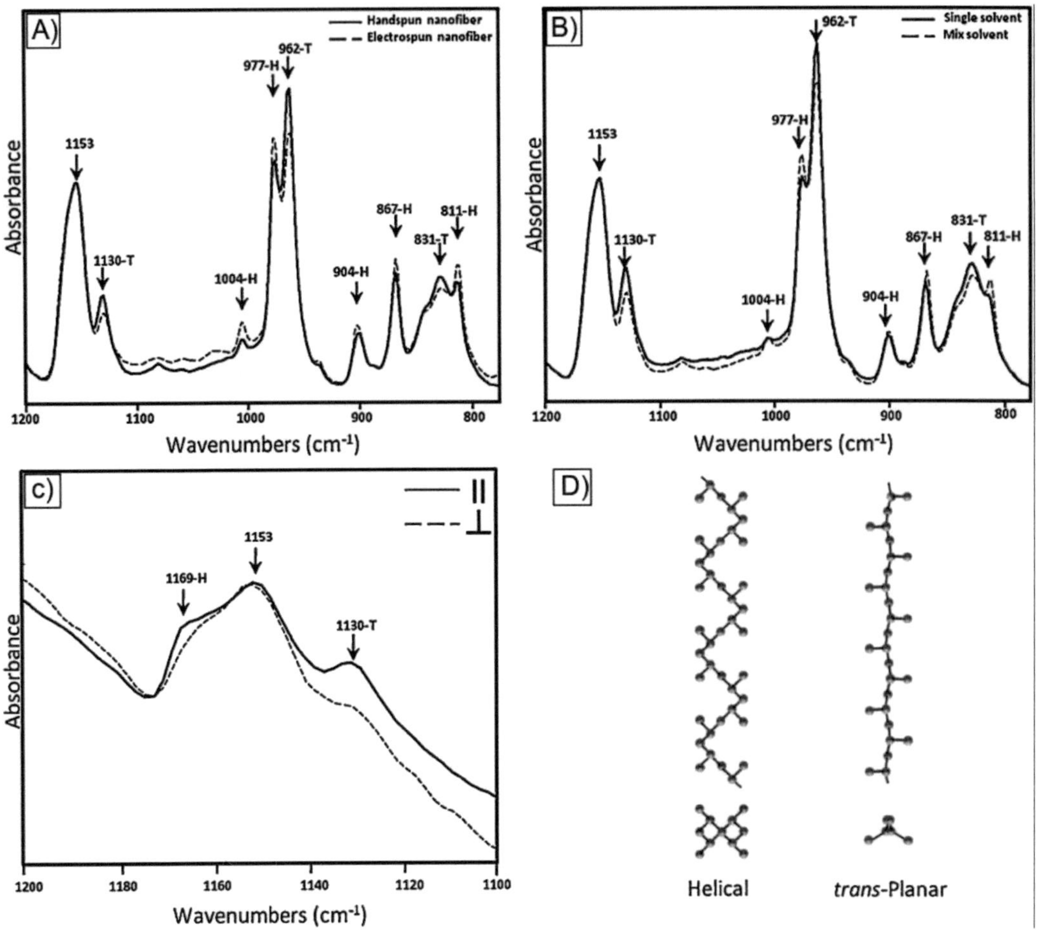

Figure 4. (a) FT-IR attenuation total reflectance (ATR) spectra of handspun and electrospun PP nanofibers. (b) FT-IR ATR spectra of handspun PP nanofibers obtained from single-solvent (cyclohexane) and mixed-solvent system (cyclohexane/acetone/DMF = 80/10/10 wt%). T: transplanar, H: helical. (c) Polarized FT-IR spectra of handspun PP nanofibers (parallel and perpendicular to the fiber axes in the macroscopically aligned fibers). T: trans-planar, H: helical. (d) Schematic representation of helical and trans-planar conformations of PP polymer chain (Reproduced with permission from Ref. [18]. Copyright 2011, Wiley-VCH Verlag GmbH & Co. KGaA, Weinheim).

the other hands, in electrospinning process, the CNTs are not aligned compared to handspun nanofiber relatively, because the electrical treatment did not be translated to mechanical force to stretch the CNTs. Thus, this novel fabrication method, handspinning, leads to well-defined nanofibers with a parallel orientation and uniaxial alignment of the CNTs to the fibers, all of which are induced by simple mechanical force.

The handspun nanofibers exhibit the enhanced physical properties such as mechanical and thermal properties. Handspinning process produces more stretched fiber than the electrospinning process, indicating that the pulling mechanical force in the handspinning induced the alignment of the polymer chains compared with electrospun nanofiber as above mention in **Figure 4** [23]. Thus, the handspinning is beneficial in that it enhances the tensile strength of the fiber by inducing

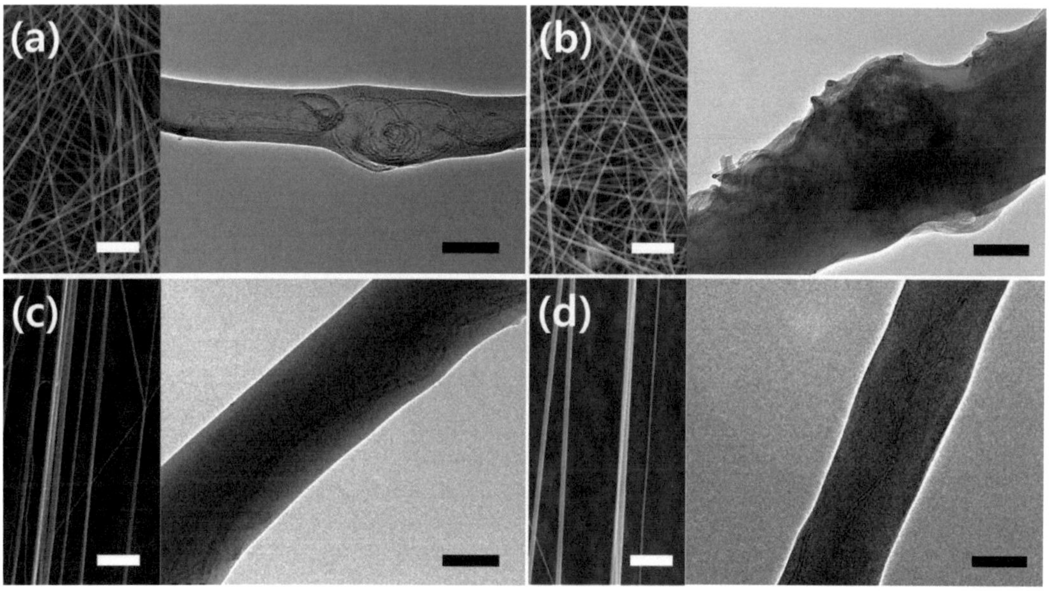

Figure 5. SEM (left) and TEM (right) images of electrospun nanofibers: (a) 0.5 wt%, (b) 1 wt% of CNTs; handspun nanofibers: (c) 0.5 wt%, (d) 1 wt% of CNTs (SEM scale bar = 5 μm, TEM scale bar = 100 nm) (Reproduced with permission from Ref. [17]. Copyright 2016, Nature Publishing Group).

polymer chain alignment. The CNTs distribution also highly affects to mechanical property of nanofiber. The tensile force was not distributed evenly along the fiber in the aggregated state; it was focused on local points where the CNT aggregates were located in the nanofiber, leading to lower tensile strength. The handspun nanofiber can achieve the evenly distributed CNTs along the fiber as observed in TEM (**Figure 5c**, **d**), thus, the handspinning method enhanced the mechanical properties of the nanofibers by a simple change in the fabrication method. Moreover, the handspun nanofibers are able to retain larger amount of CNTs. Increasing the amount of CNTs in nanofiber dramatically increased Young's modulus and tensile strength as presented in **Figure 6**. The handspinning method, in which the concentration of CNT could be increased, achieved a Young's modulus that was 1.8 times was greater than that of the electrospinning method. At the same CNT concentration for both methods, the tensile strength 2.4 times greater for the handspinning method, making it a powerful tool to attain strong nanofibers when the orientation of the CNTs is important. All related observations indicated that handspinning provides a versatile and straightforward route to obtain well-defined nanofibers. It provides a solution that allowed us to overcome the limitations in electrospinning, such as the low mechanical properties due to the limited amount of nanofillers in the polymer matrix. In addition, handspinning is universally applicable to any kinds of polymers and solvents, it enables us to create a new type of reinforced polymeric composite nanofibers with exceptional properties.

The current handspinning system also has several drawbacks. The low throughput of handspinning could be one of the drawbacks. In addition, the uniformity of nanofiber diameter relatively lows compared to conventional spinning method. Current system is presented as the proof of concept to fabricate nanofiber via simple mechanical force, it should be improved

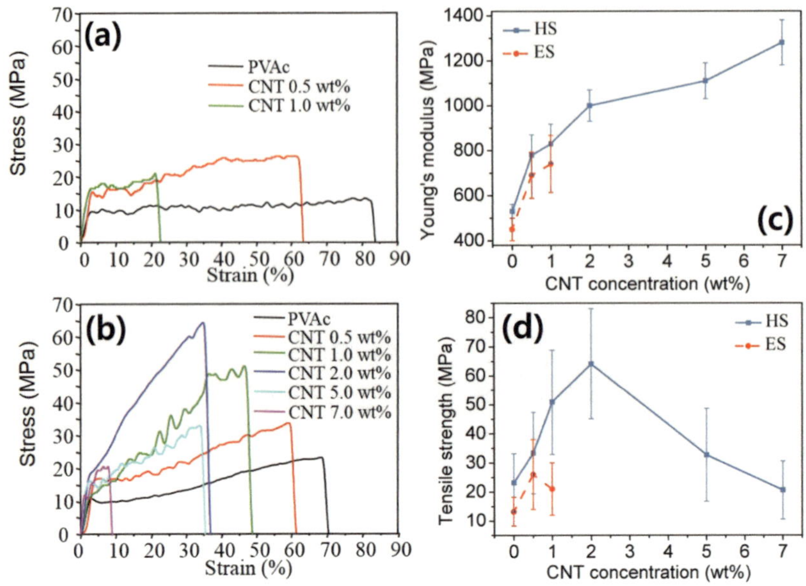

Figure 6. Stress-strain curves measured with the various CNT concentrations (a) single electrospun nanofibers; (b) single handspun nanofibers; (c and d) tensile strength and Young's modulus of single electrospun and handspun nanofibers extracted from the stress-strain curves (Reproduced with permission from Ref. [17]. Copyright 2016, Nature Publishing Group).

the instrumental set-up simultaneously achieving high efficiency as well as high quality of nanofibers. The improvement of handspinning equipment has a potential for leading to higher throughput, possibly applicable to actual application. For example: (i) In the presented set-up, pulling away speed is 40 cm/s. This indicates the length of a single fiber is 40 cm. If 100 nanofibers are produced, total length of fibers per a single operation is 40 m, indicating its potential towards comparable throughput to conventional spinning method. (ii) By making the size of plates where polymer solution is dispensed large, much larger amount of fibers can be produced. (iii) If solution dispensing is automated through the plates, time can be significantly saved. Moreover, the effect of each parameters of handspinning should be investigated, which is still not uncovered. Therefore, development of handspinning instrument is required, and the investigation of handspinning should be conducted more deeply.

2.2. Direct writing

Flexibility of electronic devices is an essential character for future smart electronic devices i.e. flexible solar cells [28], flexible displays [29], and sensor tapes [30]. Nanofibers with high conductivity as well as mechanical flexibility play a key role for constructing flexible electronic device. Especially, fabricating the ultra-long length fiber is significant in the fabrication of large-sized flexible electronics, and it also affects in manufacturing efficiency. There have been many efforts to develop the flexible device, however, precise position control and the convenient assembly process are yet to be realized in scalable production of flexible electronics.

The simplest way to fabrication of nanofiber is direct drawing from a polymer solution using a glass micropipette or an ordinary pen tip [31]. This directing writing method has been proposed by lots of literature, and they successfully constructed complex structures, patterns, and grids [32–34]. However, the drawing speed and fiber's length are too low and short to utilize in practical applications, and it fabricates the large-size macro-fiber rather than submicron or nanofiber, because drawn fibers from liquid phase are hard to bear the tensile stress. It is known that thin fibers are usually fragile and can be easily broken by large tensile stress at high drawing speed. Ondarcuhu et al. fabricated a sing fiber using a micropipette as a drawing tool [31]. In drawing method, the fiber is grown along the nucleation and precipitation of solute material in the liquid meniscus. It was an innovative and facile method to make single nanofiber, however, it was limited in draw rate less than 100 μm/s and the maximum length only reached 1 mm. Suryavanshi et al. controlled the meniscus to improve drawing speed and manage to place the fibers in well-aligned pattern [32]. However, the drawing speed was lower than 1 mm/s, which was still too low to satisfy the need for scalable mass production. Recently, Huang et al. demonstrated a simple hand-writing process to fabricate the ultralong fibers with high drawing speed as building blocks for flexible electronic devices [35]. They draw fiber from a typical pen tip by using a high molecular weight of PEO solution with sodium dodecyl sulfonate as surfactants, and added CNTs for giving a conductivity to fiber. The surfactants used for reducing the surface tension of solution, resulting that the diameter of fiber decrease down to nanosize. The simple touch and drawing process can fabricate the single nanofiber (**Figure 7a–c**). During the drawing process, the fibers diameter shrink drastically up to 60 nm due to rapid solvent evaporation. A use of typical pen instead of micropipette prevents the deformation of fiber which happens when the fiber were still in liquid phase. **Figure 7e** shows a uniform single fiber collected on glass slides, crossing 20 cm distance, and corresponding SEM image was also presented in **Figure 7f**.

This direct writing method has several benefits compared with previous reported drawing methods. The drawing speed can achieve 10 cm/s, which is 100 times much faster, and the total time needed to produce one fiber is primarily the time of suspension. It allows us to obtain the nanofiber efficiently. Moreover, the CNTs can be aligned efficiently which was observed in high resolution SEM images (**Figure 8**). As a result, thinner fibers represent higher conductivity than thicker ones and show significantly enhanced conductivity compared with isotropic CNT thin film samples. In addition, the direct written CNT/PEO fibers exhibit excellent flexibility, which is essential for flexible electronic applications.

Accurately placing a single nanofiber on desired plate is also a significant advantage of direct writing method. This method can demonstrate convenient and efficient assembly of fiber arrays, and can make flexible and conducting circuits across more than 10 cm area. They draw conducting fiber lines on the slide glass with designed pattering. Two drawn nanofibers are precisely located with desired angles of 0°, 30°, 45°, 60° and 90° (**Figure 9a**). In addition, three or four drawn fibers are perfectly cross at one points, equally dividing 360° into six and eight parts with no significant deviation (**Figure 9b-d**). It is worth noting that if two fibers are prepared in rapid sequence to create merged joints when they are not fully dried, the cross-point would be more reliable electrical connection than simple surface contact. By using multiple tips for drawing, it also achieves both fast assembly and precise control of fibers. Therefore, accurate positioning control and fast assembly have been demonstrated conveniently through simple writing method, which could be applied to flexible electronic device application with mechanical flexibility.

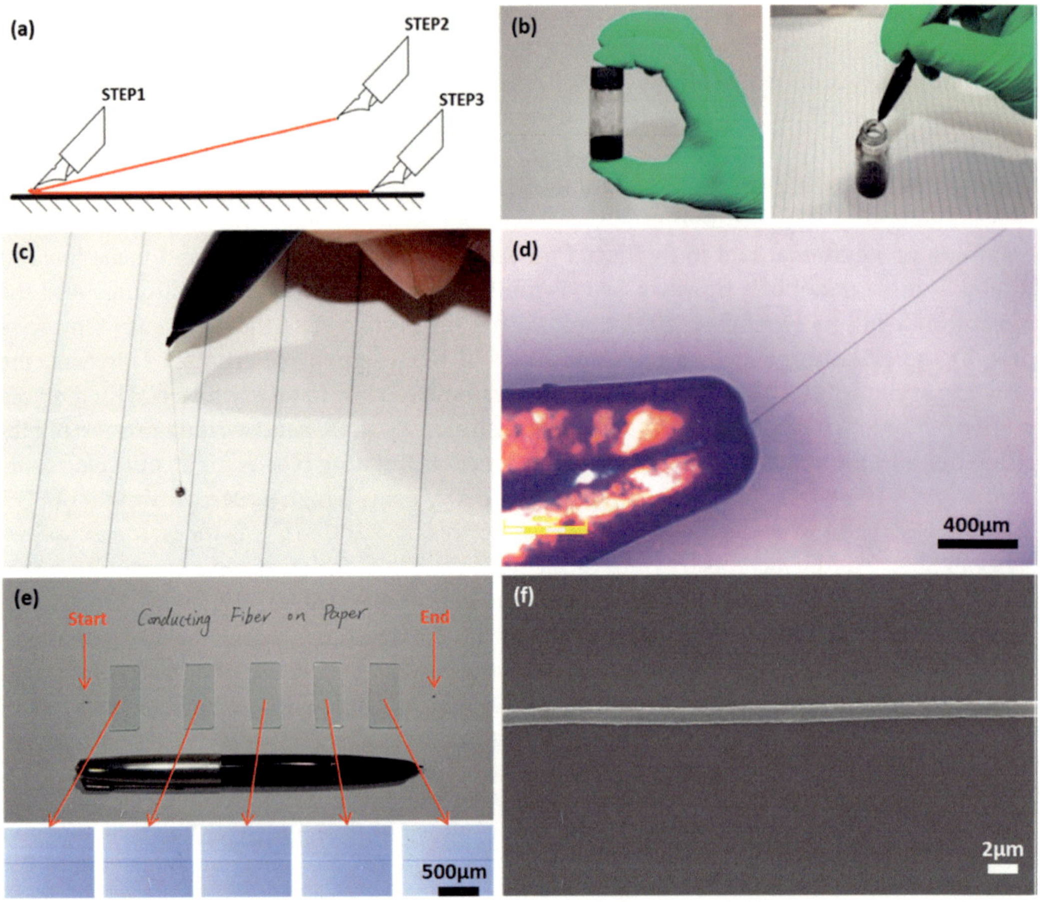

Figure 7. (a–c) A schematic and photographs of the manufacturing procedure, (d) optical microscopy of an ultralong fiber being drawn by the pen tip, (e) an ultralong and homogeneous fiber and its optical microscope images of each segment, (f) SEM image of the fiber (Adapted with permission from Ref. [29]. Copyright 2015, American Chemical Society).

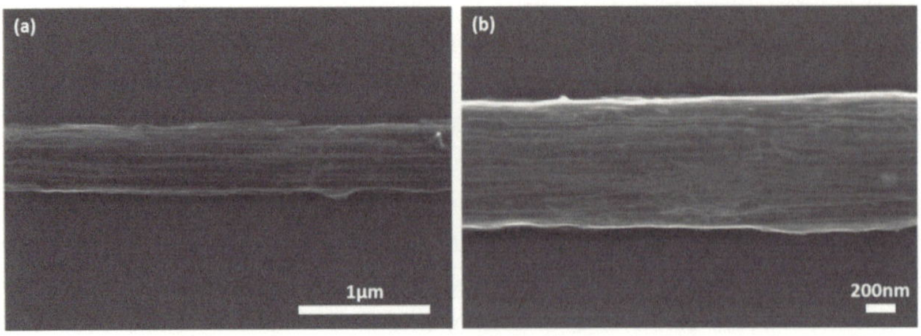

Figure 8. Confirmation of CNTs configuration in a (a) 300 nm and (b) 1 μm fiber via SEM observation (Reproduced with permission from Ref. [29]. Copyright 2015, American Chemical Society).

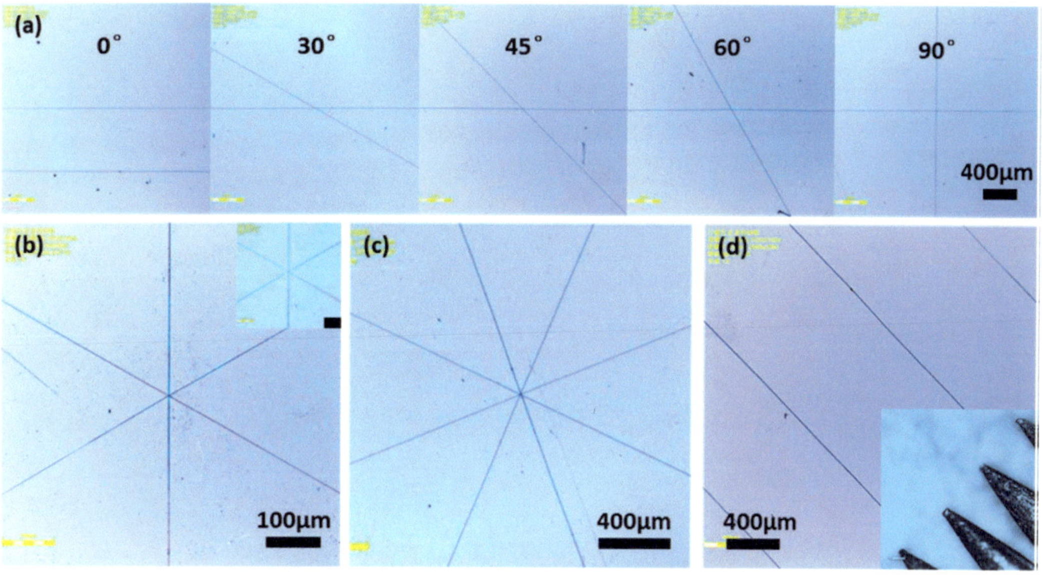

Figure 9. Demonstration of how precisely the fibers are manipulated (a) Two fibers located to form angles of 0°, 30°, 45°, 60°, and 90°; (b) three fibers that cross at one point, also observed to cross well under higher magnification (smaller figure on the top right corner, scale bar 20 μm); (c) four fibers that cross approximately at one point; (d) four parallel fibers written at one time by four aligned needle tips. (Adapted with permission from Ref. [29]. Copyright 2015, American Chemical Society).

2.3. Touch and brush spinning

The direct drawing method might be the simplest method for fabricating nanofiber. However, the drawing method has been limited in scale up and mass production, resulting that it cannot find practical applications. Recently, Tokarev et al. developed a facile and straightforward apparatus for extracting nanofibers from polymer solutions using a rotating rod or a round brush, called Touch and Brush method, which is a scalable method of nanofiber and easily fabricable 3D-scaffolds structure [36]. This method consisted with a glass rod on rotating stage (whose diameter can be chosen over a wide range of a few centimeters to more than 1 m) and a polymer solution is supplied from a needle of a syringe pump which faces the glass rod. This simple setup can be built by gluing a rod to a rotating stage from which fiber can be spun from a free-liquid surface. Simple schematic illustration was depicted in **Figure 10**. The distance between the droplet and tip of the glass rod is modulated so that the glass rod contacts the polymer droplet as it rotates. A thin filament forms after the initial touch process, and as the stage rotates, the filament stretched and fiber length increases with the diameter decreasing. The diameter of touch-spun nanofiber is modulated precisely in the range 40 nm to 5 μm by adjusting the rotational speed and polymer concentration. The fiber diameter decreases with increasing rotational speed and decreasing polymer concentration. The significant difference of the touch-spinning from other commonly used fabrication method is that it uses a simple mechanical force to manipulation of nanofiber, which is similar with handspinning method above mentioned. This simple and novel method was simply prepared with inexpensive setup that does not require special training, skills, or specialized equipment. Owing to the simplicity and ability of touch-spinning, the 3D customized scaffolds of different dimensions, shapes, mesh sizes, and fiber alignments can be easily fabricated in minutes.

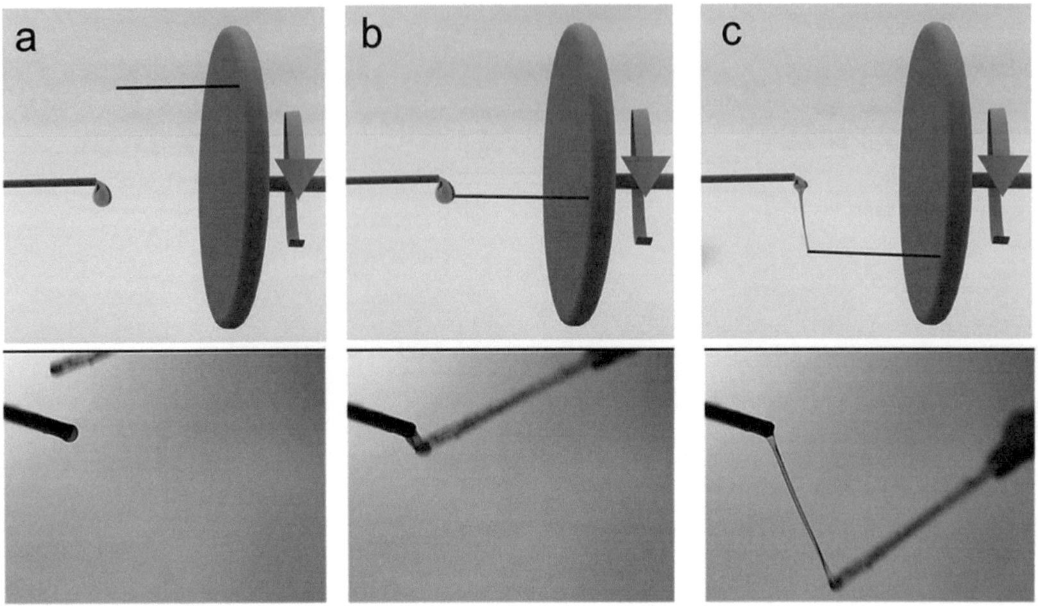

Figure 10. Schematic illustration of simple touch-spinning process a) a rotating rod on the stage and a polymer droplet, b) the rotating rod touches the droplet of the polymer solution, c) a liquid filament is formed between the rod and the tip while the rod continues rotation. (Reproduced with permission from Ref. [30]. Copyright 2015, WILEY-VCH Verlag GmbH & Co. KGaA, Weinheim).

They also realized the scalability and simplicity of touch-spinning method by using a round hairbrush composed of the order of 600 filaments instead of single rod (**Figure 11**). The brush connected and rotated to an electrical motor via brush grip, and polymer solutions supplied onto a Teflon film placed underneath the round brush. The filaments on the brush touch the polymer solution during the rotating, lots of nanofibers are spun from the free-liquid surface. From this novel method, the total length of the fibers produced by the 600-filament brush with 60 mm diameter at 3000 rpm in 5 min is 1700 km [36]. It is enough to utilize for an industrial manufacturing or typical tissue-engineering.

Moreover, this new method is a facile to wind a single filament into unidirectional, orthogonal, or randomly oriented 2D and 3D meshes with controlled density, thickness, and combinations

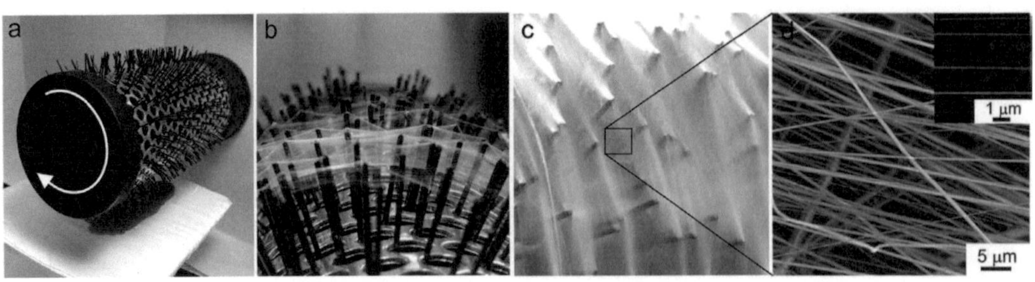

Figure 11. (a) Photographs of nanofiber fabrication process via brush-spinning method. (b) Nanofibers collected on the hairbrush rotated at 3000 rpm for 1 min and (c) for 5 min of spinning. (d) SEM image of brush-spun nanofibers (Adapted with permission from Ref. [30]. Copyright 2015, WILEY-VCH Verlag GmbH & Co. KGaA, Weinheim).

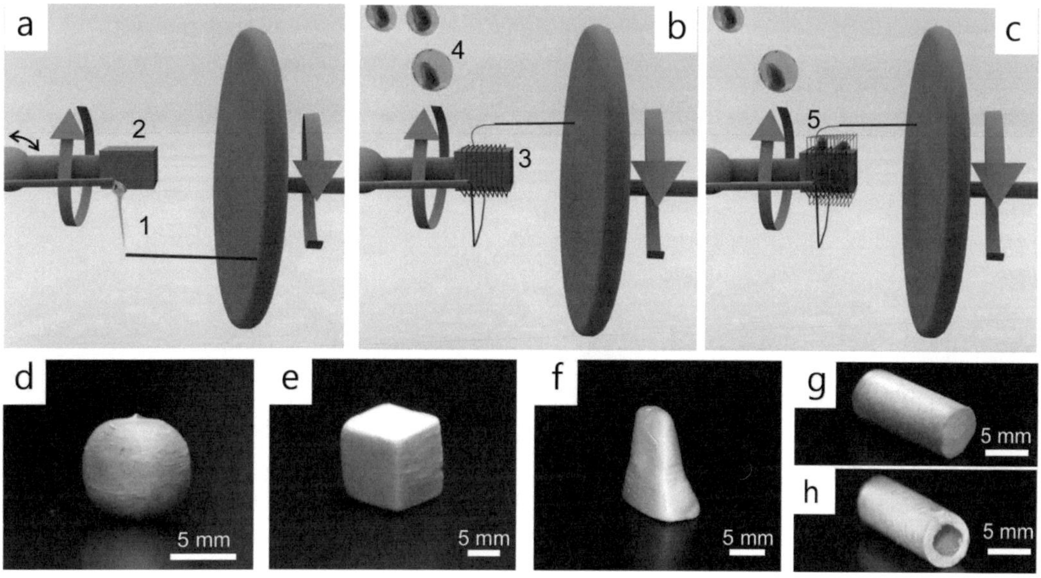

Figure 12. (a–c) Schematic illustration of the preparation of 3D scaffolds, and (d–h) different shapes and sizes of 3D scaffolds obtained by winding nanofibers (Reproduced with permission from Ref. [30]. Copyright 2015, WILEY-VCH Verlag GmbH & Co. KGaA, Weinheim).

of different fibers and materials in the scaffolds. When it considered applications of nanofiber for tissue engineering scaffolds, it could provide a very fast and practical way to produce scaffolds with controlled mesh size. Manufacturing process for 3D scaffolds was depicted in **Figure 12a–c**. After initial touch, the fiber drawn by the rod is wound onto the desired frame. The frame having complicated geometries can be easily wound because the frame can be tilted at any angle. The density of fiber on the frame can be regulated by the motion of the matrix that shuttles back and forward. From this method, different shapes and sizes of 3D scaffolds can be easily obtained by winding nanofibers onto supporting frames. Several examples of resulting matrix are presented in **Figure 12d–h**. The examples prove the capability of the touch-spinning method for fabrication of biomimetic scaffolds on different scales from macroscopic shape and dimensions to microscopic fiber dimensions and alignment into various meshes that are relevant to mesh-like structures in human tissues. It strongly implies that it could be utilized in the other various application such as filtration [37] or superhydrophobic coatings [38]. Therefore, this facile and straightforward method allows us to fabricate a productive amount of nanofiber and manufacture the 3D scaffold structure in few minute, providing an effective pathway to utilize the nanofibers.

3. Summary

Recently, there has been a lot of efforts to develop the nanofiber fabrication methods to overcome the limitations of conventional spinning method. The most widely utilized method, electrospinning, features a utilization of electrical force to fabricating nanofiber. Although

Handspinning	Direct writing	Touch and brush spinning
Facile: easy	Facile: easy	Facile: middle
Productivity: low	Productivity: low	Productivity: high
Alignment: high	Alignment: high	Alignment: low
Fiber control: unable	Fiber control: able	Fiber control: unable
Fiber uniformity: low	Fiber uniformity: middle	Fiber uniformity: low
Energy cost: low	Energy cost: low	Energy cost: low

Table 1. Comparison of handspinning, direct writing, touch and brush spinning.

electrospinning brings a highly development in the use of nanofiber, it has remained several challenges originated in the using of electrical force. The raising issues in nanofiber fabrication are the improving productivity, the fiber alignment, the enhancing the properties, and the simplicity of realization. In this chapter, we summarized the recent development of fabrication methods, handspinning, direct writing spinning, and touch/brush spinning, they provide an effective pathway to create new-types of nanofiber with outstanding properties (**Table 1**). All suggested methods use a simple mechanical stretching force, as it does not require an electrical force, which results in well-oriented polymer fibers along with the force direction, and avoids high cost and excessive use of energy in production, and offers a number of options for polymers and solvents because electrical properties are not relevant at all. Also, the spun nanofibers are expected to have enhanced mechanical properties due to change of chain conformation. Those method enable to design a desired pattern with high precision due to its easy construction, providing a potential for convenient and scalable fabrication of flexible electronic devices. Therefore, we certain that the development of fabricating methods will accelerate new applications, such as 3D scaffold, filtration, electrical devices, and those straightforward techniques open commercial opportunities for hundreds of ideas developed in the academic fields.

Author details

Hoik Lee, Davood Kharaghani and Ick Soo Kim*

*Address all correspondence to: kim@shinshu-u.ac.jp

Nano Fusion Technology Research Group, Division of Frontier Fibers, Institute for Fiber Engineering (IFES), Interdisciplinary Cluster for Cutting Edge Research (ICCER), Shinshu University, Tokida, Ueda, Nagano, Japan

References

[1] Lee H, Ryu J, Kim M, Im SS, Kim IS, Sohn D. Trace the polymerization induced by gamma-ray irradiated silica particles. Radiation Physics and Chemistry. 2016;**125**:160-164

[2] Lee H, Jeon Y, Lee Y, Lee SU, Takahara A, Sohn D. Thermodynamic control of diameter-modulated aluminosilicate nanotubes. Journal of Physical Chemistry C. 2014;**118**(15):8148-8152

[3] Lee H, Xu G, Kharaghani D, Nishino M, Song KH, Lee JS, Kim IS. Electrospun tri-layered zein/PVP-GO/zein nanofiber mats for providing biphasic drug release profiles. International Journal of Pharmaceutics. 2017;**531**(1):101-107

[4] Lee H, Kim M, Sohn D, Kim SH, Oh S-G, Im SS, Kim IS. Electrospun tungsten trioxide nanofibers decorated with palladium oxide nanoparticles exhibiting enhanced photocatalytic activity. RSC Advances. 2017;**7**(10):6108-6113

[5] Lee H, Phan D-N, Kim M, Sohn D, Oh S-G, Kim S, Kim I. The chemical deposition method for the decoration of palladium particles on carbon nanofibers with rapid conductivity changes. Nanomaterials. 2016;**6**(12):226

[6] Lee H, Nagaishi T, Phan D-N, Kim M, Zhang K-Q, Wei K, Kim IS. Effect of graphene incorporation in carbon nanofiber decorated with TiO_2 for photoanode applications. RSC Advances. 2017;**7**(11):6574-6582

[7] Lee H, Hun Song K, Soon Im S, Jung J-S, Jatoi AW, Kim IS. Fabrication of poly(vinyl alcohol)/cellulose nanofiber derivative from Kenaf bast fiber via electrospinning. Nanoscience and Nanotechnology Letters. 2016;**8**(2):168-172

[8] Fambri L, Pegoretti A, Fenner R, Incardona SD, Migliaresi C. Biodegradable fibres of poly(l-lactic acid) produced by melt spinning. Polymer. 1997;**38**(1):79-85

[9] Zhang M, Ogale AA. Carbon fibers from dry-spinning of acetylated softwood kraft lignin. Carbon. 2014;**69**:626-629

[10] Phillips DM, Drummy LF, Naik RR, Long HCD, Fox DM, Trulove PC, Mantz RA. Regenerated silk fiber wet spinning from an ionic liquid solution. Journal of Materials Chemistry. 2005;**15**(39):4206-4208

[11] Lee H, Koo JM, Sohn D, Kim I-S, Im SS. High thermal stability and high tensile strength terpolyester nanofibers containing biobased monomer: Fabrication and characterization. RSC Advances. 2016;**6**(46):40383-40388

[12] Anton F. Process and apparatus for preparing artificial threads. United States Patent Application. Journal. 1934, 1975504

[13] Li D, Xia Y. Electrospinning of nanofibers: Reinventing the wheel?. Advanced Materials. 2004;**16**(14):1151-1170

[14] Deitzel JM, Kleinmeyer JD, Hirvonen JK, Beck Tan NC. Controlled deposition of electrospun poly(ethylene oxide) fibers. Polymer. 2001;**42**(19):8163-8170

[15] Tanase M, Bauer LA, Hultgren A, Silevitch DM, Sun L, Reich DH, Searson PC, Meyer GJ. Magnetic alignment of fluorescent nanowires. Nano Letters. 2001;**1**(3):155-158

[16] Theron A, Zussman E, Yarin AL. Electrostatic field-assisted alignment of electrospun nanofibres. Nanotechnology. 2001;**12**(3):384

[17] Ellison CJ, Phatak A, Giles DW, Macosko CW, Bates FS. Melt blown nanofibers: Fiber diameter distributions and onset of fiber breakup. Polymer. 2007;**48**(11):3306-3316

[18] Zhou H, Green TB, Joo YL. The thermal effects on electrospinning of polylactic acid melts. Polymer. 2006;**47**(21):7497-7505

[19] Podgórski A, Bałazy A, Gradoń L. Application of nanofibers to improve the filtration efficiency of the most penetrating aerosol particles in fibrous filters. Chemical Engineering Science. 2006;**61**(20):6804-6815

[20] Ohzawa Y, Nagano Y, Matsuo T. Studies on dry spinning. I. Fundamental equations. Journal of Applied Polymer Science. 1969;**13**(2):257-283

[21] Horáček I, Kalíšek V. Continuous dry spinning–hot drawing preparation of fibers. Journal of Applied Polymer Science. 1994;**54**(11):1751-1757

[22] Cong H-P, Ren X-C, Wang P, Yu S-H. Wet-spinning assembly of continuous, neat, and macroscopic graphene fibers. Scientific Reports. 2012;**2**:613

[23] Lee H, Watanabe K, Kim M, Gopiraman M, Song K-H, Lee JS, Kim IS. Handspinning enabled highly concentrated carbon nanotubes with controlled orientation in nanofibers. Scientific Reports. 2016;**6**:37590

[24] Watanabe K, Kim BS, Enomoto Y, Kim IS. Fabrication of uniaxially aligned poly (propylene) nanofibers via handspinning. Macromolecular Materials and Engineering. 2011;**296**(6):568-573

[25] Chen Q, Dai L, Gao M, Huang S, Mau A. Plasma activation of carbon nanotubes for chemical modification. The Journal of Physical Chemistry. B. 2001;**105**(3):618-622

[26] Spinks GM, Shin SR, Wallace GG, Whitten PG, Kim SI, Kim SJ. Mechanical properties of chitosan/CNT microfibers obtained with improved dispersion. Sensors and Actuators B: Chemical. 2006;**115**(2):678-684

[27] Han T, Yarin AL, Reneker DH. Viscoelastic electrospun jets: Initial stresses and elongational rheometry. Polymer. 2008;**49**(6):1651-1658

[28] Haining C, Liqun Z, Huicong L, Weiping L. Growth of ZnO nanowires on fibers for one-dimensional flexible quantum dot-sensitized solar cells. Nanotechnology. 2012;**23**(7):075402

[29] Lee J, Lee P, Lee H, Lee D, Lee SS, Ko SH. Very long Ag nanowire synthesis and its application in a highly transparent, conductive and flexible metal electrode touch panel. Nanoscale. 2012;**4**(20):6408-6414

[30] Wang X, Drew C, Lee S-H, Senecal KJ, Kumar J, Samuelson LA. Electrospun nanofibrous membranes for highly sensitive optical sensors. Nano Letters. 2002;**2**(11):1273-1275

[31] Ondarçuhu T, Joachim C. Drawing a single nanofibre over hundreds of microns. EPL (Europhysics Letters). 1998;**42**(2):215

[32] Suryavanshi AP, Hu J, Yu MF. Meniscus-controlled continuous fabrication of arrays and rolls of extremely long micro- and nano-fibers. Advanced Materials. 2008;**20**(4):793-796

[33] Hu J, Yu M-F. Meniscus-confined three-dimensional electrodeposition for direct writing of wire bonds. Science. 2010;**329**(5989):313-316

[34] Nain AS, Sitti M, Jacobson A, Kowalewski T, Amon C. Dry spinning based spinneret based tunable engineered parameters (STEP) technique for controlled and aligned deposition of polymeric nanofibers. Macromolecular Rapid Communications. 2009;**30**(16):1406-1412

[35] Huang S, Zhao C, Pan W, Cui Y, Wu H. Direct writing of half-meter long CNT based fiber for flexible electronics. Nano Letters. 2015;**15**(3):1609-1614

[36] Tokarev A, Asheghali D, Griffiths IM, Trotsenko O, Gruzd A, Lin X, Stone HA, Minko S. Touch- and brush-spinning of nanofibers. Advanced Materials. 2015;**27**(41):6526-6532

[37] Qin X-H, Wang S-Y. Filtration properties of electrospinning nanofibers. Journal of Applied Polymer Science. 2006;**102**(2):1285-1290

[38] Tuteja A, Choi W, Ma M, Mabry JM, Mazzella SA, Rutledge GC, McKinley GH, Cohen RE. Designing superoleophobic surfaces. Science. 2007;**318**(5856):1618-1622

Chapter 2

Nanofiber Filaments Fabricated by a Liquid-Bath Electrospinning Method

Long Tian, Tao Yan, Jie Li and Zhijuan Pan

Additional information is available at the end of the chapter

http://dx.doi.org/10.5772/intechopen.75197

Abstract

In order to investigate the forming process of multi-needle liquid-bath electrospun nanofiber filaments, nanofiber filaments were prepared using the multi-needle liquid-bath electrospinning method in this chapter. The effect of auxiliary electrode on jet state, and bundling and drawing processes of nanofibers were studied. The results show that the forming process of nanofiber filaments was mainly influenced by electrostatic field interference, bundling process, and drawing process, including two processes: forming process of as-spun nanofiber filaments and post-drawing process. In the forming process of as-spun nanofiber filaments, when the auxiliary electrode was added, the electrostatic field interference between needles reduced, inducing the decrease of jet offsets and the enhancement of Taylor cone and jet stability, and nanofibers with skin-core structure were finally deposited on the bath in good condition. The bundling process of nanofiber filament was divided into three processes: wet process, wet-dry process, and dry process; the structure transformation of nanofiber filaments mainly occurred in the wet process. In the post-drawing process, the crystallinity and alignment degree of nanofibers increased, and nanofiber diameter decreased. The initial modulus and breaking stress of filaments increased while the breaking strain of filaments decreased. Finally, nanofiber filaments were produced with better structures and properties.

Keywords: multi-needle, electrospinning, nanofiber filaments, forming process

1. Introduction

Nanofibers, characterized by a high surface-to-volume ratio and great flexibility [1, 2], have shown wide application in areas such as biomedical engineering, filtration, and electronic engineering [3, 4]. Electrospinning is an efficient technique to produce polymeric nanofibers [5, 6]. Most electrospun nanofibers are collected in the form of nonwoven webs [7]. However,

IntechOpen

© 2018 The Author(s). Licensee IntechOpen. This chapter is distributed under the terms of the Creative Commons Attribution License (http://creativecommons.org/licenses/by/3.0), which permits unrestricted use, distribution, and reproduction in any medium, provided the original work is properly cited.

nanofiber webs are deficient in mechanical strength due to the random orientations of the nanofibers [8, 9], which limits their application in areas such as artificial organs and protective clothing [10, 11]. Nanofiber filaments or yarns are expected to be one of the best approaches to solve this problem and uncover new opportunities for the development of 3D nanofibrous structures [10, 12]. However, problems such as lack of production continuity and low mechanical strength still exist in current nanofiber filaments development.

In recent years, many methods have been developed to directly electrospin nanofibers into nanofiber filaments or yarns [12], such as using the self-assembly [13], the dual electrodes [14], the air assistant twisting device [15], the liquid bath collector [16, 17] and the rotary intermediate collecting device [11, 12, 18]. Among them, liquid-bath electrospinning method is regarded as one of the most efficient ways to produce nanofiber filaments, due to its better stability, continuity and applicability. Smit [16], Khil [19], and Pan et al. [20, 21] have successfully electrospun nanofiber filaments continuously by the liquid-bath electrospinning technique. However, there is no record about the forming process and mechanism of nanofiber filaments fabricated by liquid-bath electrospinning method.

Therefore, in this chapter, nanofiber filaments were prepared by liquid-bath electrospinning method with multi-needle. Then, the forming and post-drawing process of nanofiber filaments were studied, and the effects of electrostatic field interference, bundling process and drawing process on forming process of nanofiber filaments were emphatically discussed. Nanofiber filaments with better structures and properties could be achieved through this study. Meanwhile, this study could lay the theoretical and experimental foundation of the continuous manufacture of nanofiber filaments fabricated by liquid-bath electrospinning method with multi-needle.

2. Experimental

2.1. Materials

Pure PA6 pellets (product number 181110) were obtained from Sigma-Aldrich, and 88-wt% formic acid, purchased from Shanghai Chemical Reagent Co., Ltd., was chosen as the solvent. PA6 solutions were prepared by dissolving PA6 pellets into the solvent with continuous stirring at room temperature. The concentration of the PA6 solution was 25 wt%.

The bath solution was prepared by dissolving Peregal O (Jiangsu Jiafeng Chemical Co. Ltd.) into deionized water at room temperature, and the concentration of Peregal O solution was 0.5 wt% [21].

2.2. Filament manufacture

A homemade multi-needle liquid bath electrospinning device with an auxiliary electrode was used to continuously prepare the PA6 nanofiber filaments as in **Figure 1**. The PA6 solution was loaded into a syringe and then fed to the spinneret by the syringe pump. Both the spinneret

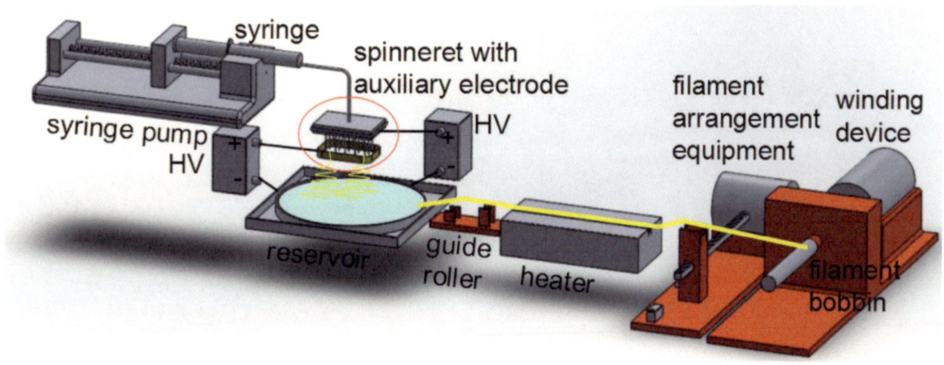

Figure 1. Schematic diagram of continuous and stable fabrication process of as-spun nanofiber filament by multi-needle electrospinning device with liquid-bath collector.

Voltage	Flow rate	Vertical distance between the needle tip and the bath surface	Drying device	
			Length	Temperature
26 kV	1.5 mL/h	60 mm	200 mm	350°C
Rotation mandrel		Auxiliary electrode		
Diameter	Rotation speed	Size (length × width)	Height	Voltage
78 mm	588 m/h	100 mm × 80 mm	17.5 mm	22 kV

Table 1. Parameters of multi-needle electrospinning process with liquid-bath collector.

and the auxiliary electrode were connected to the cathodes of high-power supplies by copper wires, while the anodes were inserted into the bottom of the bath reservoir. The as-formed high electrostatic field turned PA6 solution into nanofibers on the surface of the bath, which were initially assembled into a nanofiber filament in the bath with the guidance of a glass rod, and the filament then went through the guide roller, the heater and the filament arrangement equipment successively, before being wound on the filament bobbin in the rotation device. The as-spun nanofiber filament was continuously manufactured by the drawing force caused by the rotation of the filament bobbin [22]. The parameters of the electrospinning are listed in **Table 1**.

2.3. Post-drawing

The post-drawing process of as-spun PA6 nanofiber filaments was shown in **Figure 2**. As-spun nanofiber filaments were unwound from the unwinding roller (outer diameter = 8.2 mm) with a certain speed (v_W), then nanofiber filaments passed through a bath (0.5% Peregal O solution) for being swelled, and next traversed the drying device to obtain enough energy for nanofiber movement. Finally, nanofiber filaments were evenly wound (the speed of v_u) on an winding roller (outer diameter = 8.2 mm).

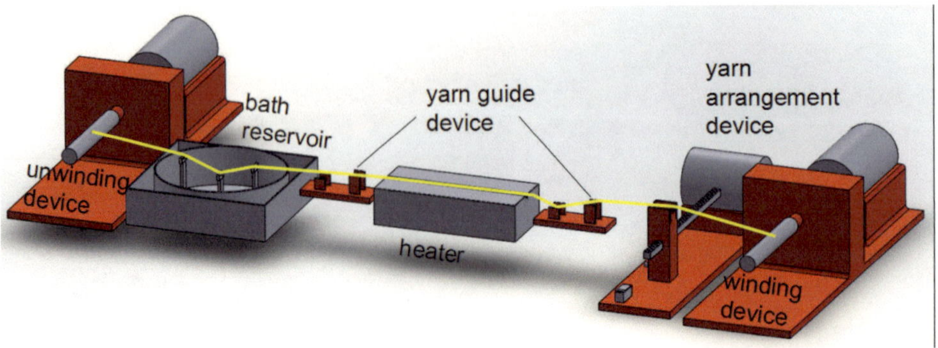

Figure 2. Schematic diagram of post-drawing device.

The post-drawing ratio (v_{draw}) could be determined by the speed ratio of winding roller and unwinding roller, as shown in Eq. 1.

$$v_{draw} = \frac{v_w}{v_u} \tag{1}$$

where v_w was the speed of winding roller, v_u was the speed of unwinding roller. Both parameters had the same unit of rotation per minute (rpm).

2.4. Characterization

The morphology of the nanofiber filaments was characterized by a scanning electron microscope (SEM) (Hitachi S-4800, Japan).

The diameters of the nanofibers in each yarn were determined by averaging 100 measurements of the nanofibers in the SEM images of the yarn using image analysis software (Image Pro Plus 5.0). Ten tensile-tested samples (length = 50 mm) were randomly selected from each tested filament. The diameter of each sample was obtained by averaging 10 measurements using a CU-2 fiber fineness tester. The diameter of each filament was obtained by averaging 10 samples' diameters.

The alignment degree (AD) of nanofiber filaments was characterized by the area ratio of aligned nanofibers and the yarn.

The mechanical properties of nanofiber filaments include initial modulus, breaking stress and strain. The measuring method was as following: first, 10 tensile-tested samples (length = 50 mm) were randomly selected from each tested filament and were maintained under standard conditions for 24 h before testing. The diameter of each sample (d_i) was obtained using a CU-2 fiber fineness tester. Second, the mechanical properties of each sample were then measured using an Instron 3365. The test parameters were set to a gauge length of 10 mm, crosshead speed of 10 mm/min, initial tension of 0.1 cN, and strength and elongation resolutions of 0.01 cN and 0.01 mm, respectively. Finally, the breaking strength (F_i), breaking length (l_i), and the strength with 1% extension (m_i) of each sample could be obtained. Therefore, the initial

http://dx.doi.org/10.5772/intechopen.75197

modulus (M), breaking stress (δ), and strain (E) of each nanofiber filament were calculated according to Eqs. 2, 3, and 4 from the average of 10 samples:

$$M = \sum_{i=1}^{10} \frac{40m_i}{\pi d_i^2} \tag{2}$$

$$\delta = \sum_{i=1}^{10} \frac{2F_i}{5\pi d_i^2} \tag{3}$$

$$E = \sum_{i=1}^{10} \frac{l_i - 10}{100} \times 100\% \tag{4}$$

3. Results and discussion

3.1. Forming process of nanofiber filaments fabricated by liquid-bath electrospinning method with multi-needle

Forming process of nanofiber filaments fabricated by liquid-bath electrospinning method with multi-needle was as following: nanofibers were electrospun using polymer solution and collected on the bath, then experienced forming process based on wet spinning, and finally bundled into nanofiber filaments by tensile force. There exist electrostatic field interferences among multiple needles during electrospinning, which could disrupt the electrospinning process, leading to the uncontinuous manufacturing of nanofiber filaments. The best way to solve this problem was to add an auxiliary electrode to multi-needle. Therefore, the stable and continuous manufacture of nanofiber filaments by liquid-bath electrospinning method with multi-needle included four processes: (1) electrospinning, (2) weakening of electrostatic field interferences among multi-needle, (3) forming of nanofibers based on wet-spinning, and (4) bundling of nanofibers.

3.1.1. Electrospinning process

In recent years, many researchers have explored and built relatively complete theoretical system of electrospinning, based on physical theories of high-voltage electrostatic field and polymer solution. In high-voltage electrostatic field, there are three processes turning polymer solution into nanofibers: (1) forming of Taylor cone, (2) straight jet, and (3) bending instability, as shown in **Figure 3**.

3.1.2. Weakening process of electrostatic field interferences among multi-needle

Electrostatic field interferences could affect electrospinning and bundling process, by affecting the electrostatic field distribution and electrospinning jets. Therefore, an auxiliary electrode was needed to weaken electrostatic field interferences.

Higher electrostatic field interferences increased the instability of Taylor cones and jets, and jet whipping during the electrospinning process got more complex [23]. These two behaviors

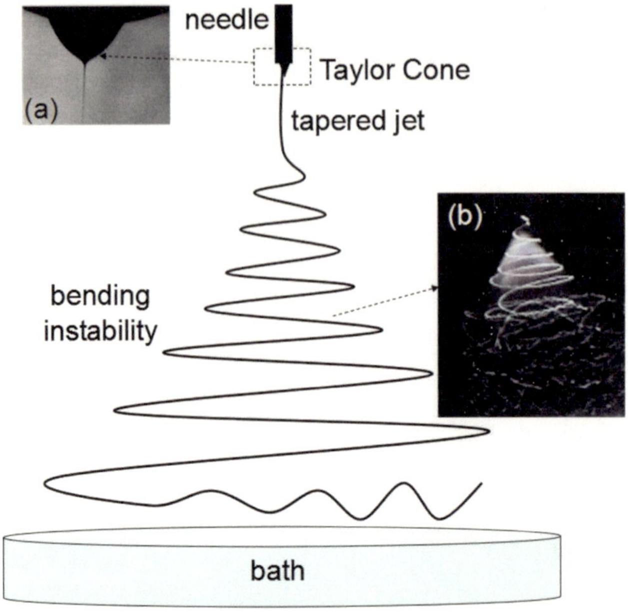

Figure 3. Schematic diagram of electrospinning process, (a) Taylor cone and straight segment, (b) bending instability.

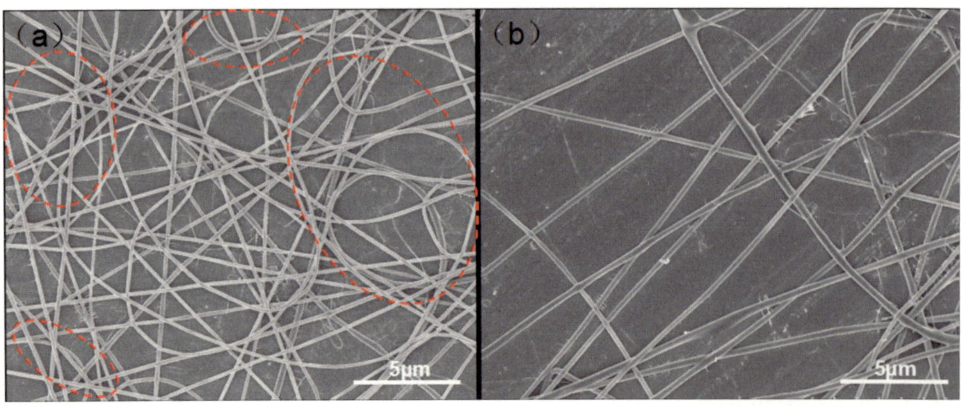

Figure 4. SEM images of nanofibers collected at the place which has a distance of 10 mm to needle tips during electrospinning process with and without rounded rectangular auxiliary electrode, (a) without auxiliary electrode, (b) with RAE.

would induce more irregular bends in the smooth nanofibers during flying, thereby leading to more intertwined and bent nanofibers, as shown in **Figure 4a**.

Higher electrostatic field interferences could also weaken electrostatic field intensity in the electrospinning space, thus resulting in the increase of nanofiber diameter due to weakened tensile force being applied to jets. Meanwhile, the offsets of jets (as shown in **Figure 5a**, 41.96° is the offset angle of the central needle in the left row) became larger, leading to larger distance

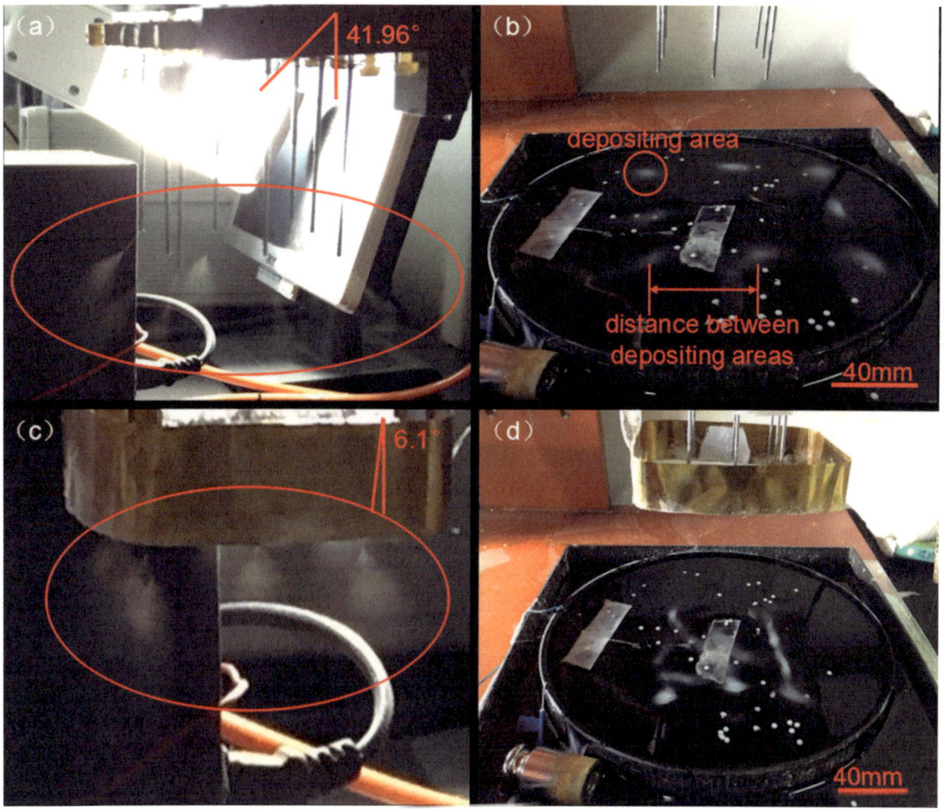

Figure 5. Effect of electrostatic field interference on electrospinning process, (a) jets state with higher electrostatic field interference, (b) nanofibers depositing state with higher electrostatic field interference, (c) jets state with lower electrostatic field interference, (d) nanofibers depositing state with lower electrostatic field interference.

between nanofiber depositing areas (the average distance between adjacent nanofiber depositing areas was 57.1 mm), which would affect bundling process of nanofiber filaments.

In this chapter, a round-corner rectangular auxiliary electrode (RAE) was applied during electrospinning process to weaken electrostatic field interferences. When electrostatic field interferences decreased, the instability of Taylor cones and jets was weakened. Meanwhile, the jet whipping during the electrospinning process also weakened [23]. Therefore, irregular bends in the smooth nanofibers during flying got less, thus resulting in less intertwined and bent nanofibers (as shown in **Figure 4b**).

With the decrease of electrostatic field interferences, the descending range of electrostatic field intensity decreased, resulting in smaller nanofiber diameter. Meanwhile, the offsets of jets decreased (as shown in **Figure 5b**, 6.1° is the offset angle of the central needle in the left row), inducing the decrease of distance between nanofiber depositing areas (the average distance between adjacent nanofiber depositing areas was 35.5 mm). Therefore, bundling process of nanofiber filament would be more stable and continuous.

3.1.3. Forming process of nanofibers based on wet spinning

Liquid-bath electrospinning nanofibers were collected in the bath, which is quite similar to coagulation bath during wet spinning. Therefore, whether this type of nanofibers had the typical structure of wet-spun fiber, skin core structure, was a subject worth exploring.

Nanofiber yarns (consisting of 12 nanofiber filaments) were overtwisted (3000 tpm, as shown in **Figure 6a**), thus the nanofibers became so stretched that they could be destroyed in this process, especially on their surfaces (as shown in **Figure 6b**). In order to explore the structures of the cracks, a slight tensile load was added to the nanofiber yarn, and the results were shown in **Figure 4c**. The structures of the surface and inner part of the nanofibers were different. This result indicated that the nanofibers had a skin core structure, which was also indicated by the cross section of the nanofibers in **Figure 4d**. This structure could be explained by the synthetic process of the nanofibers. The solvent evaporated from the nanofibers during the electrospinning before the flying nanofibers were collected in the bath, while there were still some solvent remaining in the fibers. Because the concentration of formic acid in the nanofibers was higher than that in the bath but the concentration of H_2O in the bath was higher than that in the nanofibers, counterdiffusion of these two solvents occurred in the surface of the deposited nanofibers when the nanofibers were deposited in the bath, thus forming a skin layer. Because the structure of the nanofiber skin was compact and dense, it hindered further counterdiffusion from occurring. The structure of the inner regions was therefore still loose and flexible, which lead to the skin core structure of the as-spun nanofibers [24–27].

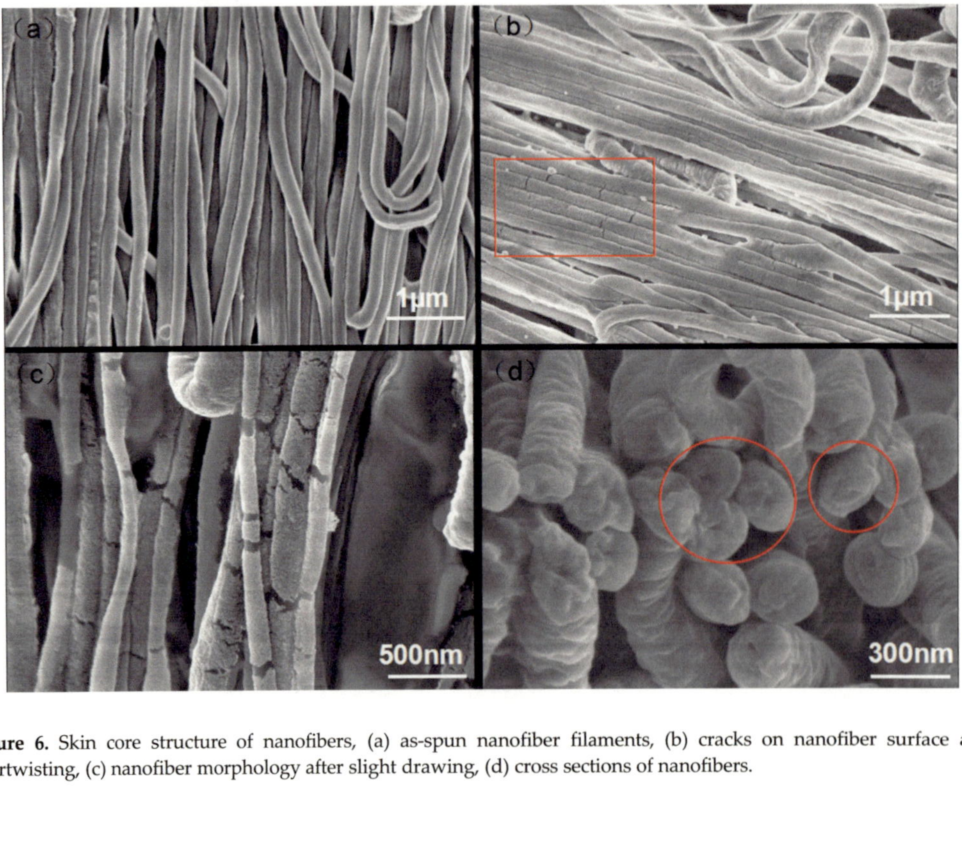

Figure 6. Skin core structure of nanofibers, (a) as-spun nanofiber filaments, (b) cracks on nanofiber surface after overtwisting, (c) nanofiber morphology after slight drawing, (d) cross sections of nanofibers.

3.1.4. Bundling process

The actual bundling process (as shown in **Figure 7a** and **b**) could be summarized as follows: electrospinning nanofibers were deposited on the bath in the form of lamelliform fiber assembly, then bundled under tensile force of winding roller and resistance force of bath, and next formed into a bundling triangular zone (as shown in **Figure 7c**) at the edge of bath reservoir turning the larger nanofiber bundle into a smaller wet one. The wet bundle then passed through the drying device becoming a dry bundle, and was finally wound on the winding roller forming the as-spun nanofiber filament. Therefore, the bundling process, as shown in **Figure 7d**, could be divided into three parts: wet process (as shown in **Figure 7a**), wet-dry process, and dry process (as shown in **Figure 7b**). Wet process consisted of transformation stage from lamelliform fiber assembly to bundling triangular zone and transformation stage from bundling triangular zone to wet bundle. Dry process consisted of transformation stage from dry bundle to as-spun filament.

3.1.4.1. Wet process

3.1.4.1.1. Transformation stage from lamelliform fiber assembly to bundling triangular zone

The nanofiber aggregative states of every process are shown in **Figure 8**. Nanofibers are randomly arranged in the lamelliform fiber assembly (as shown in **Figure 8a**), then moved into bundling triangular zone under the combined action of tensile force, water resistance force and interaction force between fibers. During this process, AD of the nanofiber bundle increased rapidly, but nanofiber diameter nearly kept the same, as shown in **Figure 9**. This is

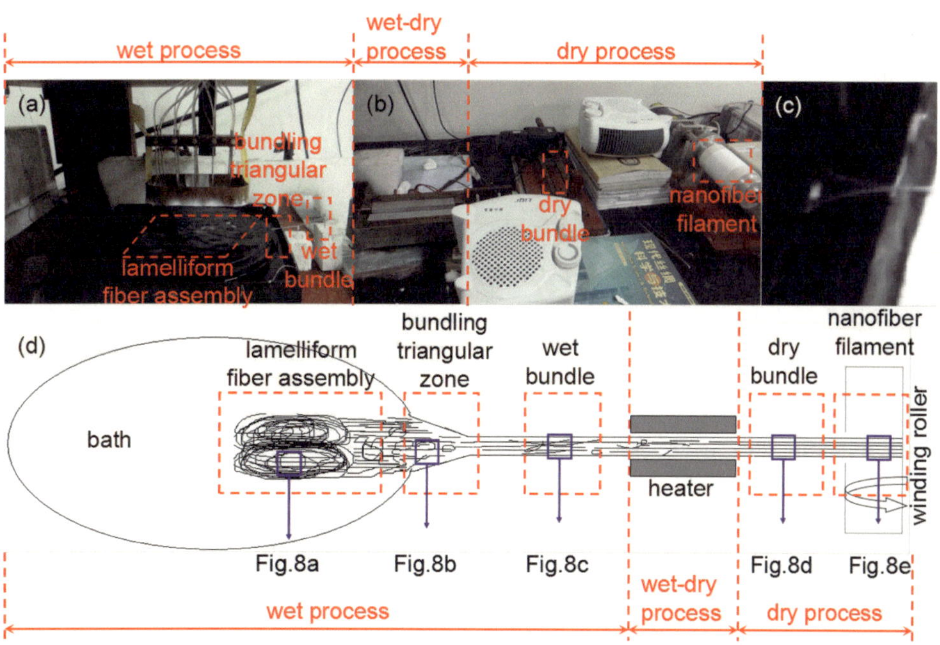

Figure 7. Bundling process of nanofiber filaments, (a) wet process, (b) wet-dry process, (c) bundling triangular zone, (d) schematic diagram of bundling process.

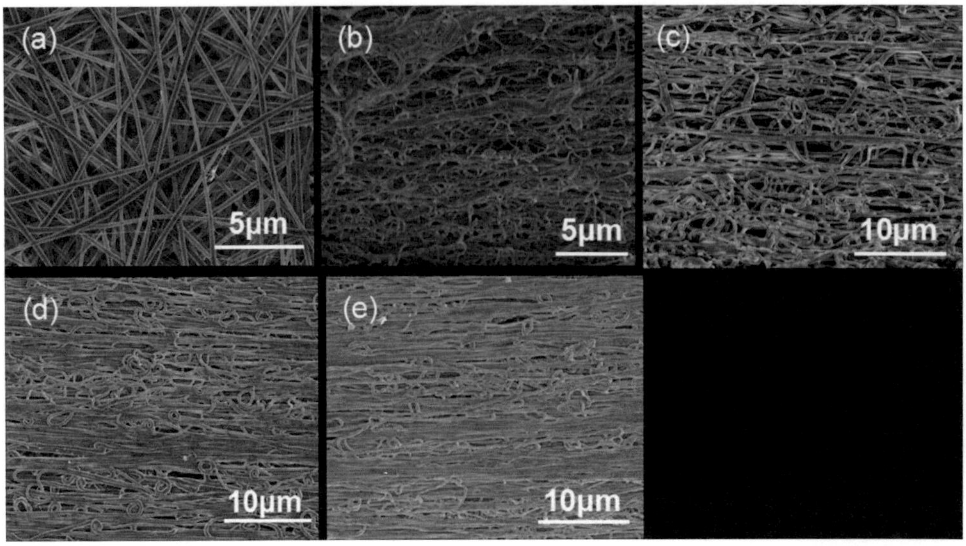

Figure 8. SEM images of nanofibers bundling morphologies during bundling process, (a) the lamelliform fiber assembly, (b) the bundling triangular zone, (c) the wet filament, (d) the dry filament, (e) the as-spun nanofiber filament.

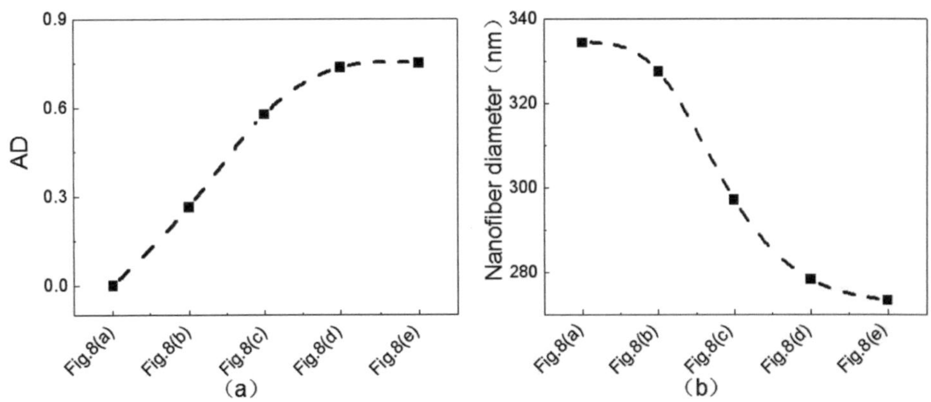

Figure 9. ADs and nanofiber diameters in nanofibers bundling states during bundling process, (a) ADs, (b) nanofiber diameters.

because the combined action was mainly contribute to the aligned movement of nanofibers, only minor combined force was used for nanofiber stretching. Although AD of nanofibers in bundling triangular zone was improved, the nonaligned nanofiber, like bent and hooked nanofibers, still occupied a dominant position, as shown in **Figure 8b**.

3.1.4.1.2. Transformation stage from bundling triangular zone to wet bundle process

Nanofibers moved forward from the bundling triangular zone to the wet bundle. During this process, a large number of nanofibers entered into a narrow space from a relatively large space, which increased the interaction force between nanofibers rapidly. Therefore, nanofibers were immediately directionally arranged, and many bent and hooked nanofibers became straight (as shown in **Figure 8c**) under the combined force of tensile force, resistance force and interaction force between fibers; thus, AD of the wet bundle was enhanced sharply (as shown in **Figure 9a**). Meanwhile, nanofibers were stretched by the combined force, so the diameter of nanofibers decreased rapidly, as shown in **Figure 9b**.

3.1.4.2. Wet-dry process

Nanofibers continuously moved forward and passed through the dry device, forming a dry bundle. Before traversing the dry device, nanofibers in wet state could still be directionally arranged, straightened and stretched by the combined force. After drying, nanofibers' position and state were nearly fixed, thus the deformability of nanofibers was weakened. Therefore, both the increase of AD and the decrease of nanofiber diameter during this process were smaller than those during the previous process, as shown in **Figures 8d** and **9**.

3.1.4.3. Dry process

The dry bundle moved on and finally wound on the winding roller. As nanofibers were fixed in the bundle, the combined force could nearly make nanofibers be further aligned or stretched. Therefore, AD and diameter of nanofibers in the as-spun filament were nearly the same with those in the dry bundle, as shown in **Figures 8e** and **9**.

To sum up, the structure transformation of as-spun nanofiber filament, including the increase of AD and the decrease of nanofiber diameter, mostly took place during wet process.

Therefore, the forming mechanism of as-spun nanofiber filament could be summarized in **Figure 10**. Polymer solution was formed into drops on the tips of multiple needles by the syringe pump. In the high-voltage electrostatic field, the electrostatic field force applied to the drops could overcome their surface tension, so jets were ejected from the drops (Taylor cones). The jets flied towards the bath in the form of straight line initially, and then in the form of multistage open loop by the combined action of viscoelasticity, surface tension, coulomb force, electrostatic field force and multiple instability. By using the auxiliary electrode, the offsets of jets became smaller, the stability of Taylor cones and jets got better, and finally the jets turned into nanofibers depositing on the bath in better form. Then the lamelliform nanofiber assembly, formed based on wet spinning, began to be bundled by external force. It was bundled into wet bundle via bundling triangular zone at the edge of the bath reservoir, and then turned into dry bundle after drying, and finally formed into as-spun filament after winding.

3.2. Post-drawing process of as-spun nanofiber filaments

The structure and bundling state of nanofibers in filaments as well as the mechanical properties of nanofibers were unsatisfactory. The most efficient way to improve the structures and mechanical properties of filaments was to use post-drawing method. When an as-spun filament was drawn along its axis under a relatively high temperature (higher than the glass transition temperature of PA6), PA6 molecular chains could move and deform along its axis [28, 29], thus the internal structure, such as orientation and crystallinity degree, could be changed; meanwhile, nanofibers could also slip and deform along its axis, so the bundling

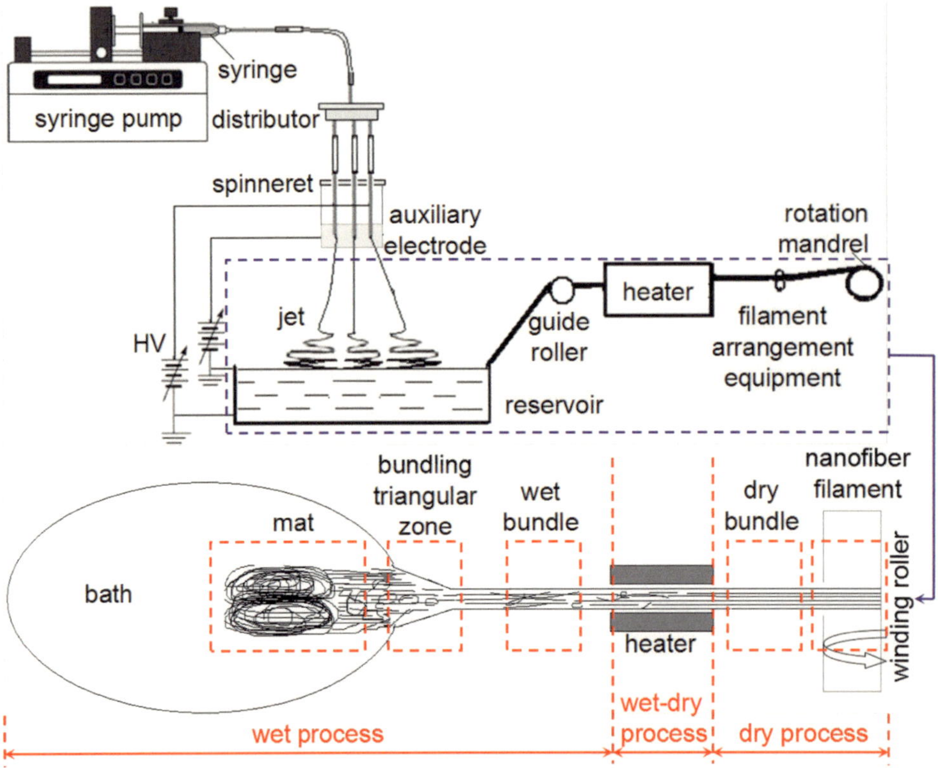

Figure 10. Schematic diagram of continuous and stable fabrication process by multi-needle electrospinning device using liquid-bath collector.

state of nanofibers in filaments could be improved. However, when the post-drawing ratio was very high, filaments were broken easily, which affected the continuous manufacture of filaments. Therefore, post-drawing ratios were selected as 1.2 times, 1.4 times and 1.6 times, to research the effect of post-drawing on structures and mechanical properties of filaments.

3.2.1. Effect of post-drawing on bundling states of filaments

Figure 11 shows bundling states of post-drawn filaments using different drawing ratios. With the increase of drawing ratios, the bent and hooked nanofibers reduced obviously, the structures of filaments got more compact, nanofiber diameter presented a decreasing trend, and ADs of filaments increased (as shown in **Figure 12**). Under a larger tensile force of post-drawing, more bent and hooked nanofibers could be straightened and obtain more plastic deformation [30], which makes nanofibers better aligned and thinner. Meanwhile, the larger tensile force could lead to more inter-fiber slippage, resulting in larger frictional force between the adjacent nanofibers, which helped more bent and hooked nanofibers to be straightened and stretched along the filament axis. Therefore, when post-drawing ratios increased, nanofiber diameters decreased, nanofiber bundling states were improved, and AD increased.

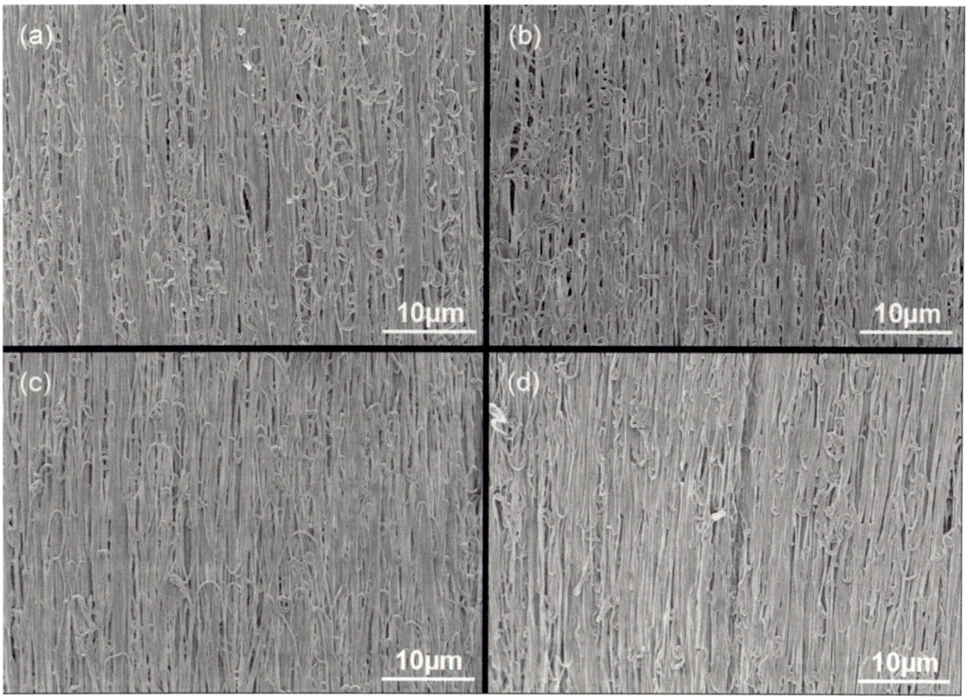

Figure 11. SEM images of nanofiber filaments post-drawn with different ratios, (a) without post-drawing (1.0 time), (b) 1.2 times, (c) 1.4 times, (d) 1.6 times.

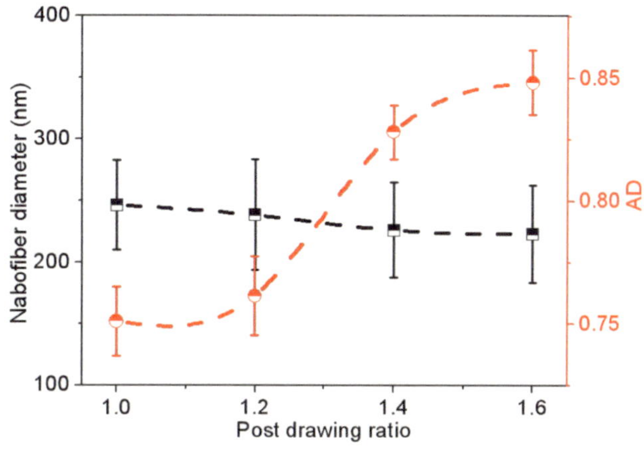

Figure 12. Nanofiber diameters and ADs of post-drawn filaments with different ratios.

3.2.2. *Effect of post-drawing on crystalline structures of filaments*

Figure 13 shows the crystalline structures of post-drawn filaments with different drawing ratios. It could be clearly observed that all post-drawn filaments exhibited the same diffraction peaks at the 2θ angles of 20.5°, 21.5° and 24.5° in **Figure 13a**. 20.5° and 24.5° represent α crystal and are

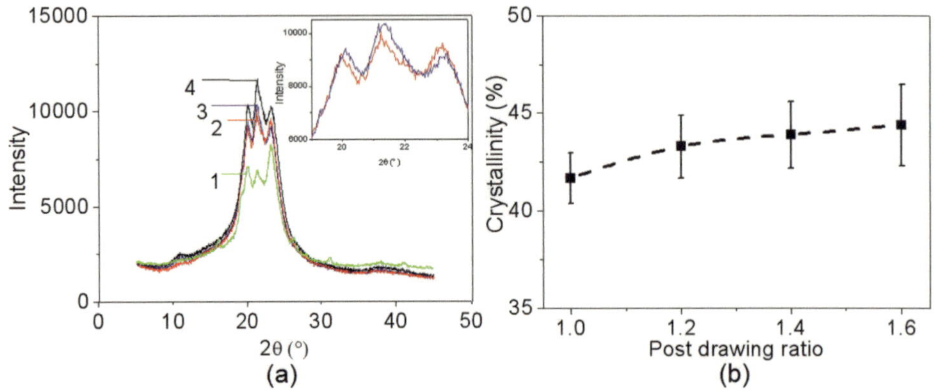

Figure 13. Crystallinities of nanofibers in filaments post-drawn with different ratios, (a) XRD curves, (b) crystallinities.

indexed as (200) and (002)/(202) reflections, respectively; while 21.5° represents γ crystal and are indexed as (001) reflections [31, 32]. This behavior indicated that post-drawing process did not change the crystalline structure of the filaments. However, the intensity of the diffraction peak of filaments became stronger with the increase of drawing ratios, especially at the 2θ angle of 21.6°, which indicated increase in crystallinities of filaments (as shown in **Figure 13b**). When post-drawing ratios increased, which meant larger tensile force was applied to filaments, more polymeric molecular chains in nanofibers could be more easily arranged regularly [33, 34], the distance between adjacent molecular chains decreased at the same time, so more hydrogen bonds between molecular chains could be formed in nanofibers, indicating the amount of γ crystal increased [35, 36]. In addition, some γ crystal could transform into α crystal under tensile force [37, 38]. Therefore, with the increase of drawing ratios, the crystallinities of filaments increased.

3.2.3. *Effect of post-drawing on mechanical properties of filaments*

Figure 14 shows stress–strain curves and main mechanical property indexes of post-drawn filament with different drawing ratios. With the increase of drawing ratios, the initial modulus and breaking stress of filaments were both improved, while the breaking strain of filaments reduced. Those variation trends were codetermined by mechanical properties of nanofibers, interaction between nanofibers and AD of filaments. With increasing drawing ratios, the nanofiber crystallinities increased, which was beneficial to promoting the initial modulus and breaking stress of filaments, the breaking strain was reduced accordingly. For filaments, the effect of filament bundling state was dominated by filament mechanical properties. With increasing the drawing ratios, more inter-fiber slippages happened, leading to the decrease of inter-fiber gaps, thus structures of filaments became more compact, inducing the enhancement of cohesive force between nanofibers; meanwhile, more oblique, bent and hooked nanofibers were straightened and directionally arranged. Those behaviors all contributed to the increase of initial modulus and breaking stress, and the decrease of the breaking strain.

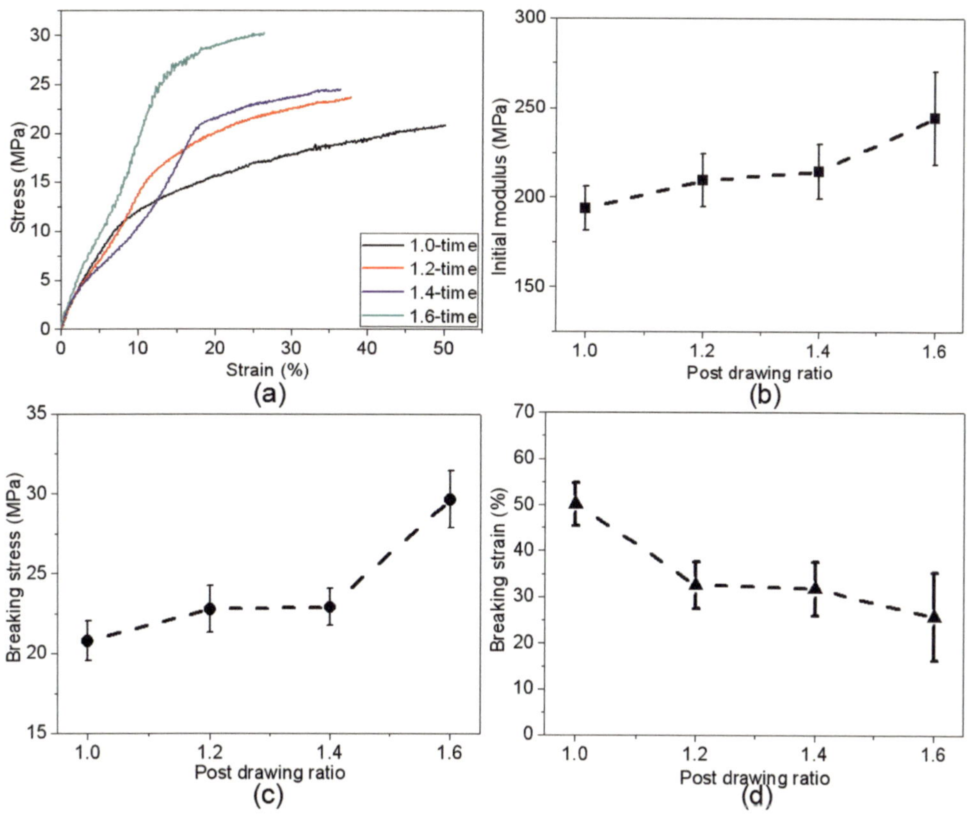

Figure 14. Mechanical properties of filaments post-drawn with different ratios, (a) stress–strain curves, (b) initial modulus, (c) breaking stress, (d) breaking strain.

4. Conclusion

The forming process of nanofiber filaments included two processes: forming process of as-spun nanofiber filaments and post-drawing process. In the forming process of as-spun nanofiber filaments, when the auxiliary electrode was added, the electrostatic field interference between needles reduced, inducing the decrease of jet offsets and the enhancement of Taylor cone and jet stability, and nanofibers with skin-core structure were finally deposited on the bath in good condition. The bundling process of nanofiber filament was divided into three processes: wet process, wet-dry process and dry process; the structure transformation of nanofiber filaments, including the increase of ADs and the decrease of nanofiber diameters, mainly occurred in the wet process. In the post-drawing process, the crystallinity and AD of nanofibers increased, and nanofiber diameter decreased. The initial modulus and breaking stress of filaments increased while the breaking strain of filaments decreased. Finally, nanofiber filaments were produced with better structures and properties.

Acknowledgements

Financial support for this work was provided by the Nantong Science and Technology Project (GY12016025).

Author details

Long Tian[1,2], Tao Yan[1], Jie Li[3] and Zhijuan Pan[1,2,4]*

*Address all correspondence to: zhjpan@suda.edu.cn

1 College of Textile and Clothing Engineering, Soochow University, Suzhou, China

2 Nantong Textile and Silk Industrial Technology Research Institute, Nantong, China

3 Jiangsu Textiles Quality Services Inspection Testing Institute, Nanjing, China

4 National Engineering Laboratory for Modern Silk, Soochow University, Suzhou, China

References

[1] Arinstein A, Zussman E. Electrospun polymer nanofibers: Mechanical and thermodynamic perspectives. Journal of Polymer Science Party B: Polymer Physics. 2011;**49**:691-707

[2] Del Gaudio C, Ercolani E, Nanni F, Bianco A. Assessment of poly (ε-caprolactone)/poly (3-hydroxybutyrate-co-3-hydroxyvalerate) blends processed by solvent casting and electrospinning. Materials Science and Engineering: A. 2011;**528**(3):1764-1772

[3] Su CI, Lai TC, Lu CH, Liu YS, Wu SP. Yarn formation of nanofibers prepared using electrospinning. Fibers and Polymers. 2013;**14**(4):542-549

[4] Fang J, Lin T, Tian W, Sharma A, Wang X. Toughened electrospun nanofibers from crosslinked elastomer-thermoplastic blends. Journal of Applied Polymer Science. 2007; **105**(4):2321-2326

[5] Konwarh R, Karak N, Misra M. Electrospun cellulose acetate nanofibers: The present status and gamut of biotechnological applications. Biotechnology Advances. 2013;**31**: 421-437

[6] Tungprapa S, Puangparn T, Weerasombut M, Jangchud I, Fakum P, Semongkhol S, et al. Electrospun cellulose acetate fibers: Effect of solvent system on morphology and fiber diameter. Cellulose. 2007;**14**(6):563-575

[7] Shuakat MN, Lin T. Recent developments in electrospinning of nanofiber yarns. Journal of Nanoscience and Nanotechnology. 2014;**14**:1389-1408

[8] Wu SH, Qin XH. Uniaxially aligned polyacrylonitrile nanofiber yarns prepared by a novel modified electrospinning method. Materials Letter. 2013;**106**:204-207

[9] Niu H, Gao W, Lin T, Wang X, Kong L. Composite yarns fabricated from continuous needleless electrospun nanofibers. Polymer Engineering & Science. 2014;**54**(7):1495-1502

[10] Hajiani F, Jeddi AAA, Gharehaghaji AA. An investigation on the effects of twist on geometry of the electrospinning triangle and polyamide 66 nanofiber yarn strength. Fibers and Polymer. 2012;**13**:244-252

[11] Li N, Hui Q, Xue H, Xiong J. Electrospun polyacrylonitrile nanofiber yarn prepared by funnel-shape collector. Materials Letter. 2012;**79**:245-247

[12] Ali U, Zhou Y, Wang X, Lin T. Direct electrospinning of highly twisted, continuous nanofiber yarns. Journal of Textile Institute. 2012;**103**:80-88

[13] El-Aufy AK. Nanofibers and Nanocomposites Poly(3,4-Ethylene Dioxythiophene)/Poly (Styrene Sulfonate) by Electrospinning [Thesis]. Philadelphia: Drexel University; 2004

[14] Pan H, Li LM. Continuous aligned polymer fibers produced by a modified electrospinning method. Polymer. 2006;**47**:4901-4904

[15] Ko F, Gogotsi Y, Ali A, Naguib N, Ye H, Yang GL, et al. Electrospinning of continuous carbon nanotube-filled nanofiber yarns. Advanced Materials. 2003;**15**(14):1161-1165

[16] Smit E, Buttner U, Sanderson RD. Continuous yarns from electrospun fibers. Polymer. 2005;**46**:2419-2423

[17] Teo WE, Gopal R, Ramaseshan R, Fujihara K, Ramakrishna S. A dynamic liquid support system for continuous electrospun yarn fabrication. Polymer. 2007;**48**(12):3400-3405

[18] Dabirian F, Ravandi SH, Sanatgar RH, Hinestroza JP. Manufacturing of twisted continuous PAN nanofiber yarn by electrospinning process. Fibers and Polymers. 2011;**12**(5):610-615

[19] Khil MS, Bhattarai SR, Kim HY, Kim SZ, Lee KH. Novel fabricated matrix via electrospinning for tissue engineering. Journal of Biomedical Materials Research Part B: Applied Biomaterials. 2005;**72**(1):117-124

[20] Liu Y, Li J, Pan ZJ. Effect of spinning conditions on the mechanical properties of PA6/MWNTs nanofiber. Journal of Polymer Research. 2011;**18**(6):2055-2060

[21] Li J, Tian L, Pan N, Pan ZJ. Mechanical and electrical properties of the PA6/SWNTs nanofiber yarn by electrospinning. Polymer Engineering & Science. 2014;**54**(7):1618-1624

[22] Tian L, Zhao C, Pan Z. Fabrication of high-alignment-degree nanofiber filaments using multi-needle electrospinning equipment with an auxiliary electrode. Science of Advanced Materials. 2015;**7**(11):2327-2335

[23] Karatay O, Dogan M, Uyar T, Cokeliler D, Kocum IC. An alternative electrospinning approach with varying electric field for 2-D-aligned nanofibers. IEEE Transactions on Nanotechnology. 2014;**13**(1):101-108

[24] Ge H, Liu H, Chen J, Wang C. The skin-core structure of poly (acrylonitrile-itaconic acid) precursor fibers in wet-spinning. Journal of Applied Polymer Science. 2008;**108**(2):947-952

[25] Lv M, Ge H, Chen J. Study on the chemical structure and skin-core structure of polyacrylonitrile-based fibers during stabilization. Journal of Polymer Research. 2009;**16**: 513-517

[26] Kitagawa T, Yabuki K, Young RJ. An investigation into the relationship between processing, structure, and properties for high-modulus PBO fibers. II. Hysteresis of stress-induced Raman band shifts and peak broadening, and skin-core structure. Journal of Macromolecular Science, Part B. 2002;**41**(1):61-76

[27] Wang M, Jin HJ, Kaplan DL, Rutledge GC. Mechanical properties of electrospun silk fibers. Macromolecules. 2004;**37**:6856-6864

[28] Barkoula NM, Peijs T, Schimanski T, Loos J. Processing of single polymer composites using the concept of constrained fibers. Polymer Composites. 2005;**26**(1):114-120

[29] Rongved L, Kurjian C, Geyling F. Mechanical tempering of optical fibers. Journal of Non-Crystalline Solids. 1980;**42**(1):579-584

[30] Butler R, Prevorsek D, Kwon Y. Optimization of fiber drawing processes in terms of the filament state variables. Polymer Engineering & Science. 1982;**22**(6):329-344

[31] Zhao Z, Zheng W, Tian H, Yu W, Han D, Li B. Crystallization behaviors of secondarily quenched nylon 6. Materials Letters. 2007;**61**(3):925-928

[32] Tian L, Zhao C, Li J, Pan Z. Multi-needle, electrospun, nanofiber filaments: Effects of the needle arrangement on the nanofiber alignment degree and electrostatic field distribution. Textile Research Journal. 2015;**85**(6):621-631

[33] Pan ZJ, Liu HB, Wan QH. Morphology and mechanical property of electrospun PA 6/66 copolymer filament constructed of nanofibers. Journal of Fiber Bioengineering and Informatics. 2008;**1**(1):47-54

[34] Xie Z, Niu H, Lin T. Continuous polyacrylonitrile nanofiber yarns: Preparation and dry-drawing treatment for carbon nanofiber production. RSC Advances. 2015;**5**(20):15147-15153

[35] Kim GM, Michler GH, Ania F, Calleja FB. Temperature dependence of polymorphism in electrospun nanofibres of PA6 and PA6/clay nanocomposite. Polymer. 2007;**48**(16):4814-4823

[36] Stephens JS, Chase DB, Rabolt JF. Effect of the electrospinning process on polymer crystallization chain conformation in nylon-6 and nylon-12. Macromolecules. 2004;**37**(3):877-881

[37] Ibanes C, De Boissieu M, David L, Seguela R. High temperature behaviour of the crystalline phases in unfilled and clay-filled nylon 6 fibers. Polymer. 2006;**47**(14):5071-5079

[38] Miri V, Persyn O, Lefebvre JM, Seguela R, Stroeks A. Strain-induced disorder–order crystalline phase transition in nylon 6 and its miscible blends. Polymer. 2007;**48**(17):5080-5087

Chapter 3

Electrospinning of Collagen and Its Derivatives for Biomedical Applications

Wei Peng Lu and Yanchuan Guo

Additional information is available at the end of the chapter

http://dx.doi.org/10.5772/intechopen.73581

Abstract

Collagen, gelatin and their derived polypeptides can act as multifunctional natural polymers with excellent physicochemical properties for biomedical applications. The use of electrospinning technology can convert collagen materials into nanofibrous materials that exhibit porous micro-nanostructures with good mechanical properties and excellent biocompatibility profiles. In this chapter, a systematic review of collagen electrospinning is presented and related applications are introduced including tissue engineering (e.g., artificial skin, artificial vasculature, cartilage repair, etc.), drug delivery, hemostatic dressings, periodontal restoration, biofilms, and wound dressings will now be discussed.

Keywords: collagen, gelatin, electrospinning, fiber

1. Introduction

Electrospinning is an easy and inexpensive process that can be used to prepare nanofibers from almost any soluble or fusible polymer under the action of a high electrostatic field. These electrospun nanofibers often possess extremely high surface areas, high porosities, tunable pore structures, and superior mechanical properties, which means they can be processed into materials with a wide variety of structure and function design [1–6]. Due to these advantages, electrospun nanofibers have been used for a broad range of biomedical and industrial applications, such as protective clothing, wound dressings, drug delivery applications, and tissue engineering [7–10].

The emergence of electrospinning (also known as electrostatic spinning) originated more than 100 years ago, with Zeleny first reporting on a variant of the electrospinning process in

IntechOpen

© 2018 The Author(s). Licensee IntechOpen. This chapter is distributed under the terms of the Creative Commons Attribution License (http://creativecommons.org/licenses/by/3.0), which permits unrestricted use, distribution, and reproduction in any medium, provided the original work is properly cited. (cc) BY

1917 [11]. The first patent that described the process of electrospinning appeared in 1934, with Formalas describing apparatus to prepare polymer filaments, using an approach that relied on electrostatic repulsions between surface charges [12]. In 1966, Simons invented an electrospinning device that could be used to produce ultra-thin nonwoven fibers [13]. In 1981, Manley and Larrondo described how fabricated continuous fibers could be produced using melt-electrospinning polyethylene (PE)/polypropylene (PP) blends [14]. Then, Rutledge et al described new techniques for the production of polymer fibers by electrospinning in 1995 [15]. These breakthroughs resulted in a large increase in the number of reports describing electrospinning processes, with hundreds of different electrospinning polymers having been described for ultrafine fiber material applications. These include widely used synthetic polymers, such as polylactic acid, polyglycolide, polyethylene oxide, polycaprolactone, as well as natural polymers such as silk fibroin, fibrous protein, collagen, chitosan, hyaluronic acid, and gelatin. Natural biorenewable polymers often exhibit advantages in terms of their biocompatibility and biodegradability, making them a popular choice for a wide range of biomedical applications.

Among these natural polymers, collagen is one of the major extracellular matrix proteins that are present in many tissues and organs [16–18]. For instance, collagen in human skin accounts for approximately 70% of the extracellular matrix, where it functions as a network of elongated fibers to provide structural stability [19]. So far, more than 29 different types of collagen have been documented, with different species employing 46 different types of polypeptide chain for their assembly [20]. Collagen has many functional characteristics that are favorable for cell and tissue growth, and as a consequence it has been widely used as a biomaterial for medical and biotechnological applications [21–23].

Gelatin is a denatured protein that is obtained by acid, alkaline, and enzyme processing of collagen, which can exhibit similar physical and biological properties to those of collagen [24–28]. Due to its excellent biocompatibility, biodegradability, and immunogenicity profiles, gelatin is one of the most common biopolymers used for biomaterial applications [29–31]. The polypeptides derived from collagen play an important role in tissue remodeling [26]. Furthermore, collagen-derived peptides are also known to express biological activities, such as antioxidant, anti-osteoporosis, anti-photoaging properties, as well as acting as inhibitors of angiotensin-I converting enzyme [32, 33].

However, conventional polymeric products derived from collagen and gelatin (and their derived polypeptides) do not exhibit well-defined nanostructures, meaning that their mechanical, adhesion, and hydrophilic properties are not ideally suited for many biomedical applications. It has been shown that electrospinning is a useful technique to transform collagen, gelatin, or polypeptide into nanostructured fibers materials that can display small-size effects, high specific surface areas, and high porosities [21]. Furthermore, it also has been demonstrated that nonwoven electrospun collagen, gelatin, or polypeptide nanofibers can be used as ideal models to mimic the biochemical and ultrastructural properties of the extracellular matrix of tissue [34, 35]. In terms of performance, nanofibers materials also possess strong adsorbent powers, good filtration qualities, excellent obstruction performance, good binding affinities, and desirable moisturizing properties. Therefore, this chapter will now review the electrospinning techniques that can be used to transform collagen, gelatin, or derived polypeptides into nanofibers materials for biomedical applications.

2. Devices used for collagen electrospinning

Electrospinning is carried out using a Taylor cone that is generated by applying a high voltage to a polymer or melt solution, which results in formation of a liquid jet that is formed from interaction of a continuously increasing electric field with the surface tension of a droplet surface. During this process, a liquid jet starts oscillating to generate an irregular high-frequency spiral motion that leads to stretching of a fiber that is accompanied by fast volatilization of the solvent. This results in nano-scale fibers either being formed in a random manner on a collecting device, or being cross-linked into a membrane when a move and rotate collecting device is employed for processing [36, 37]. The electrospinning apparatus is comprised of three parts: a high-voltage direct-current power supply, a liquid supply unit, and a collecting device [38]. Depending on the nature of the liquid supply unit, the collagen electrospinning apparatus can be classified into two types: needle electrospinning or needleless electrospinning. The liquid supply unit of a needle electrospinning unit normally consists of a microinjection pump, a syringe, a single spinneret (or spinneret array), and a metal conducting wire that connects the spinneret to a high-voltage power source. The collecting device is comprised of a carrier, such as a metal plate, metal roller, metal disk, metal drum, or a ground metal conducting wire. The key component of any needle electrospinning assembly is the spinneret, which serves to prevent the monomer solution from becoming too viscous and solidifying under the spinning conditions, thus preventing blockage of the needle and potential damage to the equipment. However, the limited throughput of the needle spinneret means that the efficiency of polymer production using this technique is generally low [39]. Needleless electrospinning techniques represent a better way for high efficiency of polymer nanofiber production that can solve many of the limitations associated with traditional needle electrospinning techniques [40–42]. There are many methods for generating the surface disturbance of spinning solutions that is required for needless electrospinning, including ultrasonic disturbance, agitation, and acoustic bubble disturbance [43]. The stability of these surface disturbance processes is related to the quality (diameter distribution and L/D ratio distribution) of the needleless electrospinning membrane materials, resonance factors, liquid levels, solution concentrations, and solution viscosities. Needleless electrospinning equipment employs various types of electric field distribution, which can sometimes lead to membranes being produced with uneven thickness. Therefore, it is important that new collecting devices are designed to solve these important performance problems.

3. Introduction of composite nanofibers electrospinning technology

To date, electrospinning is the common method available to prepare nanofibers directly, quickly, and continuously under mild conditions in a low-cost, fast, and efficient manner [44, 45]. Depending on their applications, electrospinning nanofibers can be divided into two categories: single-component nanofibers and composite nanofibers [46]. Early reports described nanofibers that were prepared by electrospinning homogeneous polymers, with changes of starting materials, solution conditions, and electrospinning parameters used to prepare single-component nanofibers with different morphologies and properties. However, many of these single-component nanofibers such as collagen [47–49], gelatin [50, 51], elastin [52], and fibrinogen [53, 54], have some limitations including weak mechanical properties, poor processability, poor moisture resistance, rapid degradation rate, and potential immunogenic properties [55, 56]. Thus, composite or hybrid nanofibers with different compositions (e.g., organic/organic, organic/inorganic) have been proposed as promising materials that exhibit physicochemical properties arising from both the host and guest materials [57]. For example, Chen et al. used electrospinning to prepare collagen/chitosan nanofiber membranes, which could promote the growth of dermal and epidermal layers [58]. Gu et al. used electrospinning to prepare porous biocompatible nanofibers mats from poly(L-lactide)/gelatin, which exhibited controlled evaporative water losses and promote fluid drainage, which made them potentially useful materials for wound dressing applications [59]. For the preparation of composite nanofibers, electrospinning technics can be divided into three fundamental types: (1) blend electrospinning; (2) mixing electrospinning; and (3) coaxial electrospinning. Blend electrospinning is the most commonly used method, which involves a process whereby a spinning solution is generated by mixing different polymers in a defined ratio. Mixing electrospinning refers to an electrospinning process that employs two or more separate liquid feeding devices containing different solutions. The electrostatic field results in each polymer being stretched into nanofibers which then overlap with each other to form composite nanofiber membranes. Coaxial electrospinning involves the use of a spinneret consisting of two or more capillary tubes with different inner diameters which results in a defined gap between the two capillary tubes. The same (or different) electrostatic field is applied to the inner and outer layers of electrospun solutions which results in solutions of the core and surface polymers being expelled from each coaxial nozzle to generate a concentric stratified flow. Because each of the electrospinning solutions has a short confluence time and low diffusion coefficient, they are stretched into coaxial composite nanofibers by the presence of the electric field force. In a comparative study, Chen et al. prepared a range of composite nanofiber membranes using blend electrospinning, mixing electrospinning, and coaxial electrospinning. They found that the composite nanofibers membranes prepared by coaxial electrospinning had high regularity, the membranes produced by blend electrospinning had good moisture resistance, while nanofiber membranes fabricated by using mixing electrospinning exhibited the highest mechanical strength [60].

4. Technical factors of collagenous electrospinning

There are various factors that can influence the morphology and structural properties of nanofibers produced in the electrospinning process. These include: (1) the properties of the electrospinning solution, such as concentration, viscosity, electrical conductivity, surface tension, and distribution of polymer molecular weight; (2) process parameters, such as electric voltage, spinning temperature, spinning speed, collection speed, and spinning distance; and (3) environmental parameters, such as temperature, moisture, air velocity, and atmospheric composition.

Kazanci investigated the role of temperature, solvent, and pH on the properties of collagen-type I nanofibers that were prepared under electrospinning conditions. They found that decreasing the temperature by 10°C, the PP-II (folded) fraction ratio of the resultant fibers increased from 37 to 52.5%. Moreover, nanofibers obtained from acidic solutions contained 59% of PP-II, suggesting that the collagen structure was well preserved [61]. Dulnik et al. prepared electrospun polycaprolactone/gelatin and polycaprolactone/collagen nanofibers using various solvents (hexafluoroisopropanol and a mixture of acetic acid and formic acid). The result showed that electrospun PCL/gelatin and PCL/collagen nanofibers obtained by various solvents had similar morphologies, although there were some differences in their internal structures that affected their susceptibility toward biodegradation [62].

Lu et al. used water as a solvent to electrospin pure gelatin solution, finding that low concentrations of gelatin had low viscosity which were insufficient to produce continuous nanofibers, resulted in the formation of unwanted microbeads. When the concentration of gelatin used was >25%, the spinning solution became too viscous, which inhibited efficient electrospinning, resulted in a few nanofibers [63].

Wang et al. investigated the influence of solution concentration, salt concentration, solvent type, ambient temperature, and environmental humidity on the electrospinning of gelatin solutions. Their results showed that using polymers with high dielectric constants and solvents with low volatility solvents resulted in formation of nanofibers with small diameters. The diameters of the nanofibers were found to increase as the temperature rose. Low temperatures were not conducive to effective volatilization of solvent, leading to prolonged solidification time and the production of superfine nanofibers. The morphology of nanofibers was also found to be affected by ambient humidity. When relative humidity levels were increased to 45%, the resultant nanofiber membranes were shown to contain small areas of reticular formation and 'beads-on-a-string' structures. As the relative humidity was increased to 60%, the nanofiber membranes produced were contained uneven nanometer-diameter distributions with associated micro-bead structures. This study also revealed that the average diameter of nanofibers decreased with an increase in ambient humidity [64] (**Figure 1**).

An et al. studied the influence of formic acid concentration, gelatin concentration, and electric voltage on the electrospinning process used to prepare gelatin/polylactide composite nanofibers. They found that the use of low formic acid concentrations (50%) resulted in nanofibers, along with the formation of significant amounts of microbeads. Increasing formic acid concentration led to the formation of randomly arrayed ultrafine nanofibers, without any microbeads. For example, increasing the concentration of formic acid from 70 to 98%, resulted in an increase in the average diameter of gelatin nanofibers from 208 to 312 nm. Moreover, it was found that the concentration of gelatin affected the surface tension and viscosity of the spinning solution, with high surface tension favoring formation of microbeads, and high viscosity minimizing their formation. For example, when gelatin concentration was too low, discrete microbeads or beaded nanofibers were formed, while an increase in concentration of gelatin to >10% resulted in exclusive formation of smooth nanofibers. It was also found that the average diameter of nanofibers increased from 260 to 335 nm as the gelatin concentration rose from 10 to 23% [65]. Huang et al. had also reported that elastin-mimetic peptide polymers could be electrospun into nanofibers, who investigated the effect of polypeptide content and solution flow rate on the morphology and mechanical properties of the resultant nanofibers [66].

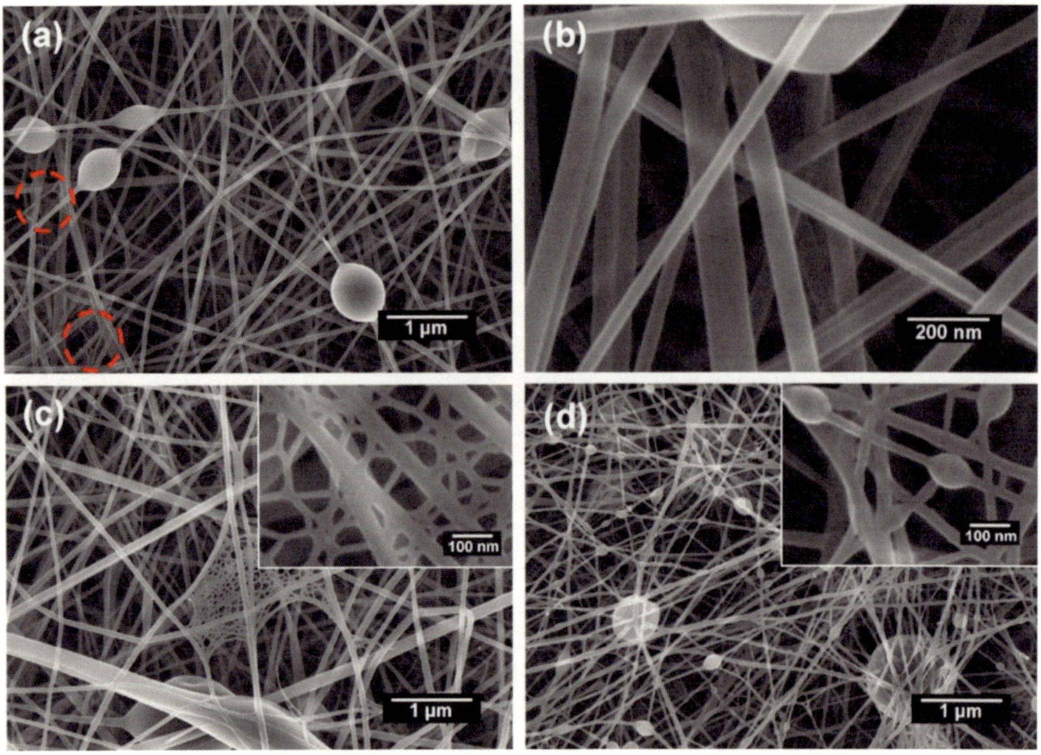

Figure 1. FE-SEM images of as-spun fibers using 10 wt.% gelatin/FA solution containing NaCl (0.1 wt.%) at temperature of 15°C (a and b) (voltage of 30 kV, and RH of 25%) and different RH: (c) 45% and (d) 60% (voltage of 30 kV, temperature of 24°C). The insets show the corresponding higher magnification images [64].

It is now clear that various factors in the electrospinning process play a crucial role in determining the 'spinnability' of a polymer solution and the physical and chemical properties of the resultant nanofibers. However, adjustment of these parameters can result in improvements to the electrospinning process, thus enabling the properties of functional nanofibers to be tuned to meet the needs for biomedical applications.

5. Applications of electrospun collagen

Electrospinning technology can be used to convert collagen materials into nanofibers materials that exhibit porous micro-nanostructures with good mechanical properties and excellent biocompatibility profiles. Potential uses of these nanofibers for biomedical applications include tissue engineering (e.g., artificial skin, artificial vasculature, cartilage repair, etc.), drug delivery, hemostatic dressings, periodontal restoration, biofilm, and wound dressings. Some applications of polymers derived from composite collagen and gelatin nanofibers, along with information on their preparation (type and solvent) and fiber diameter are provided in **Table 1**.

Composition	Solvent	Fiber diameter (nm)	Targeted applications	Ref.
Collagen/PLGA	HFP	50–500	Bone tissue scaffolds	[67]
Collagen/PHBV	HIFP	300–600	Scaffold for tissue engineering	[68]
Collagen/PCL	HFP	210–225	Vascular tissue engineering	[69]
Collagen/PLLA	HFIP	1290–1560	Tissue engineering	[70]
Collagen/TPU	HFP	700–800	Tissue engineering and functional biomaterials	[49]
Collagen/alginate/ chitosan/hydroxyapatite	Ethanol/ glycerol	300–800	Scaffold for regenerating bone tissue	[71]
Collagen/PCL	HFP	300	Implantable functional muscle tissues for patients with large muscle defects	[72]
Collagen/PCL	HFP	520	Autologous nerve grafts or proximal nerve stumps	[73]
Collagen/PLGA	HFIP	185–314	Long-term drug delivery of various pharmaceuticals	[74]
Collagen/PHBV/GO	TFE	400–500	Wound coverage material	[75]
Collagen/chitosan	HFP	434–691	Vascular and nerve tissue engineering	[76]
Collagen/collagen	HFP	210–540	Tissue engineering	[77]
Collagen/elastin	HFP	110–1120	Cardiovascular tissue engineering	[78]
Collagen/PLC	HEP	520	Vascular tissue engineering	[79]
Collagen/PLC	HEP	600–900	Human skin tissue engineering	[80]
Collagen/PEO	Aqueous	100–150	Wound dressings and tissue engineering	[81]
Collagen/PLC	HFP	500–600	Peripheral nerve regeneration	[82]
Collagen/PLLA-CL	HFP	100–200	Vascular tissue engineering	[83]
Collagen/PLLA-CL	HFP	120–520	Vascular tissue engineering	[84]
Collagen, elastin/PLGA, PCL, PLLA, or PLLA-CL	HFP	470–770	Cardiovascular tissue engineering	[85, 86]
Gelatin/PLLA-CL	TFE	50–500	Human skin tissue engineering	[87]
Gelatin/PCL	HFP TFE	640–880 50–1000	Cardiovascular tissue engineering	[78, 88]
Gelatin/PCL	TFE	2790–4630	Tissue engineering	[89]
Gelatin/PCL	TFE	160–232	Neural tissue engineering	[90]
Gelatin/zein	Acetic acid	380.3–695.5	Bioactive delivery in food industry	[91]
Gelatin/tannic Gelatin/gallic Gelatin/caffeic Gelatin/ferulic	Acetic acid	280	Biomaterials and tissue engineering Delivery system in medicinal or food industry	[92]

Composition	Solvent	Fiber diameter (nm)	Targeted applications	Ref.
Gelatin/GO	Acetic acid	200 ± 50–270 ± 50	Tissue engineering and wound dressing	[93]
Gelatin/PLC/QAS	TFE	180 ± 40–220 ± 80	Antibacterial wound dressing	[94]
Gelatin/chitosan	Acetic acid	202 ± 13.4–223.1 ± 69.8	Drug release	[95]
Gelatin/PCL	Acetic acid	250–400	Tissue engineering	[96]
Gelatin/PCL/CeNP	HFIP	616 ± 216	Wound dressing material	[97]
Gelatin/PLLA	Aqueous acetic acid solution	86–148	Wound dressing	[59]
Gelatin/PCL	TFE	800–2660	Various medical applications	[49]
Gelatin/PCL or collagen/PCL	HFIP or AA/FA	\	Wound healing, scaffolds, drugs delivery	[51]
Gelatin/NaCl	Formic acid	37–90	Filtration, tissue engineering, energy storage, sensors, and catalysis	[53]
Gelatin/PLLA	Formic acid	–	Tissue engineering scaffolds	[54]
Gelatin/PCL	TFE	312 ± 146	Guided bone regeneration	[98]
Gelatin/siloxane	Formic acid	–	Bone tissue engineering	[99]
Gelatin/PANi	HFP	61 ± 13–803 ± 121	Biocompatible scaffolds for tissue engineering	[100]
Gelatin/PLGA/FBF	TFE	310 ± 24	Drug delivery	[102]
Gelatin/PLA/PA	Formic acid	412	Healing material	[103]
Gelatin/$AgNO_3$	Glacial acetic acid/distilled water	280 ± 40	Wound dressing materials	[105]
Gelatin/PLCL/PLA	HFP	540 ± 230	Periodontal regeneration	[107]
Gelatin/PLA/MET		960 ± 560		
Gelatin/PLA/HAP		650 ± 440		
Gelatin/PCL/ZnO	HFP	56–1180	Periodontal regeneration	[108]

Table 1. Examples of collagen and gelatin electrospun nanofibers.

5.1. Tissue engineering applications

The three-dimensional microstructure of materials prepared from electrospinning collagen-derived materials can be used to effectively stimulate tissue regeneration. The network structure of these materials serves to promote the integration and recruitment of new tissue into their fibers scaffolds, thus accelerating the growth of new tissue. The main function of the nanofiber scaffold is to provide a suitable three-dimensional environment that cells can adhere to and proliferate.

Yu et al. used of electrospinning to prepare composite nanofiber scaffolds made of alginate, chitosan, hydroxyapatite, and collagen, whose porous structure was beneficial for cell infiltration and growth. What is more, the release of collagen from the scaffolds was, respectively, 17 and 2% of that from the collagen film after immersing in SBF and collagenase solution for 10 days, which indicated that the disintegration of scaffold for bone tissue engineering can be reduced comparing to conventional collagen scaffold. Therefore such a composite mat would be applicable as a scaffold for regenerating bone tissue [71] (**Figure 2**).

Choi et al. used electrospinning techniques to fabricate aligned polycaprolactone/collagen nanofiber meshes that were used to guide morphogenesis of skeletal muscle cells and enhance their cellular organization. Comparison with randomly oriented nanofibers, the results revealed that unidirectionally oriented nanofibers were better at inducing muscle cell alignment and myotubule formation [72] (**Figure 3**).

Lee et al. utilized electrospinning techniques to prepare PCL/collagen nerve conduits for complex peripheral motor nerve regeneration studies using end-to-side neurorrhaphy techniques. Results revealed that axonal continuity was normally recovered 8 weeks after surgery, with the muscle function recovery occurring after 1–20 weeks, and recovery of donor nerve function occurring after 20 weeks. Therefore, the use of electrospun PCL/collagen nerve conduits appeared to have great potential as materials for complex peripheral motor nerve repair [73] (**Figure 4**).

Composite nanofiber scaffolds have been prepared by electrospinning polycaprolactone and gelatin, which were then modified using cell-derived factor-1α. Experiments revealed that these scaffolds not only had good biocompatibility profiles, but also accelerated the healing of skull injuries in mice [98]. Ren et al. used gelatin and silicone to prepare composite nanofiber scaffolds as bone repairing material incorporating Ca^{2+} ions. Their results indicated that these composite materials could promote accumulation of bone apatite and the differentiation and

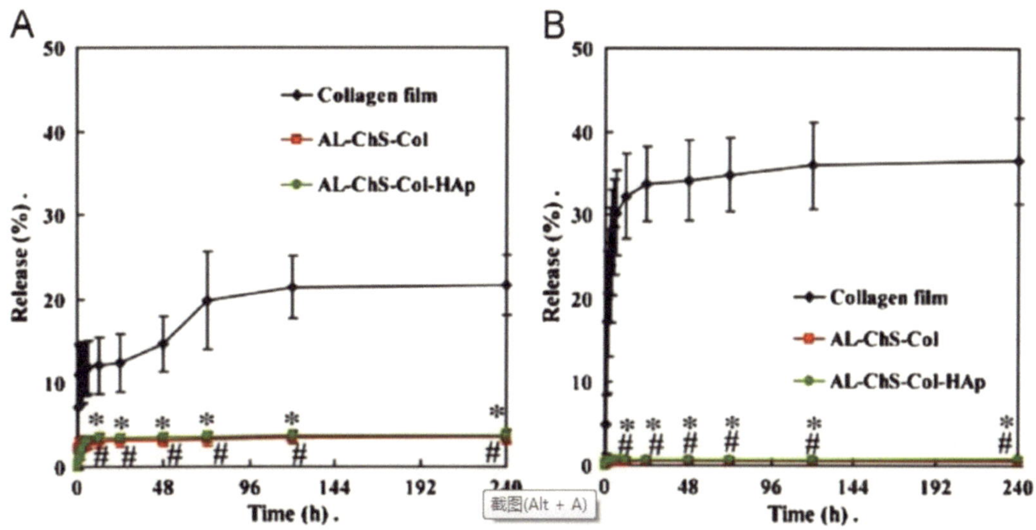

Figure 2. (A) Release of collagen in SBF solution for 10 days and (B) release of collagen in collagenase solution 10 days [71].

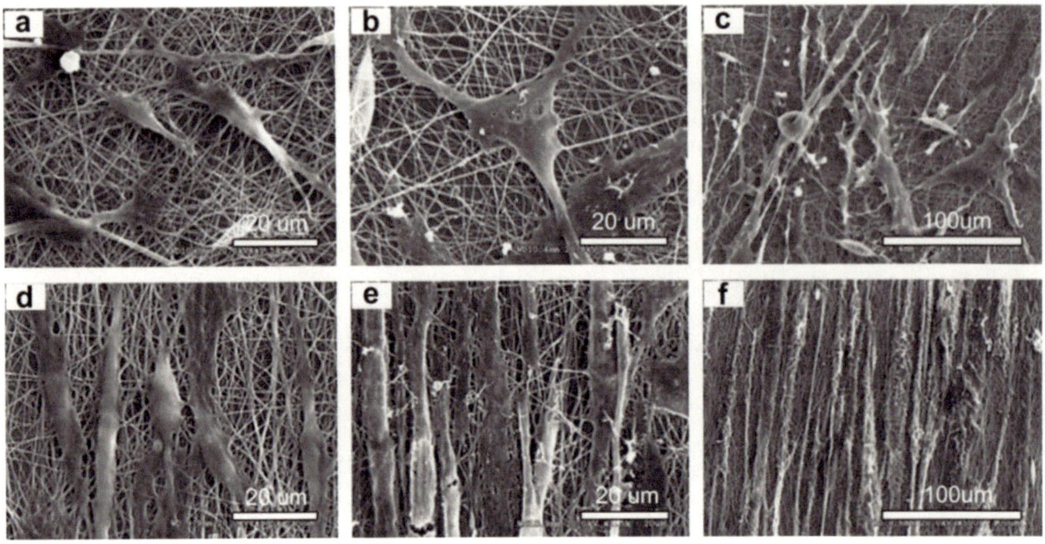

Figure 3. SEM images of human skeletal muscle cells on electrospun PCL/collagen nanofiber membranes: (a–c) randomly oriented and (d–f) aligned electrospun membranes, (a and d) 1 day and (b and e) 3 days after cell seeding and (c and f) 7 days after cell differentiation [72].

proliferation of osteoblasts [99] (**Figure 5**). Li et al. used electrospinning to prepare composite nanofiber scaffolds from polyaniline and gelatin, whose properties were dependent on the amount of polyaniline present. For example, when the amount of polyaniline was increased from 0 to 5% (w/w), the average diameter of the fibers decreased from 803 ± 121 to 61 ± 13 nm, and its elastic coefficient increased from 499 ± 207 to 1384 ± 105 MPa. These composite nanofiber scaffolds not only had excellent mechanical strength, but also had excellent biocompatibility in cell culturing experiments [100].

Blit et al. utilized the electrospinning to fabricate fibrous scaffolds which were subsequently surface modified with polypeptide and used them as membrane supports to culture smooth muscle cells (SMCs). Results showed that SMCs seeded onto these elastin-like polypeptide-4 membranes exhibited a spindle-like morphology, with good actin filament organization and smooth muscle myosin heavy chain expression. Therefore, these electrospun nanocomposite

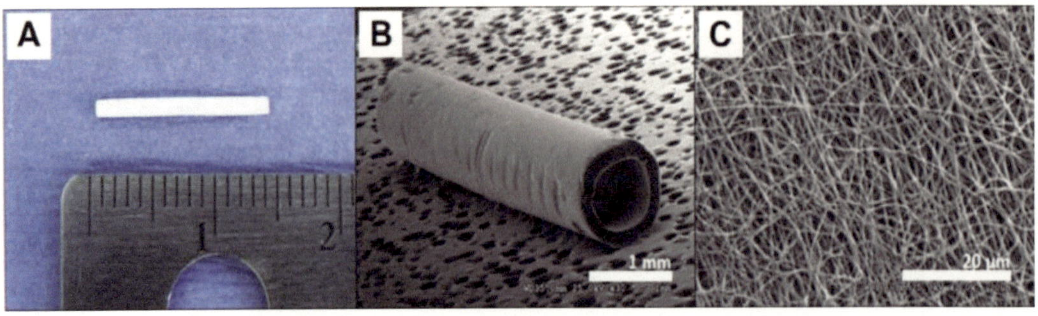

Figure 4. (A) Gross appearance and SEM images of electrospun PCL/collagen conduits: (B) entire (30×) and (C) surface (2.0× K) [73].

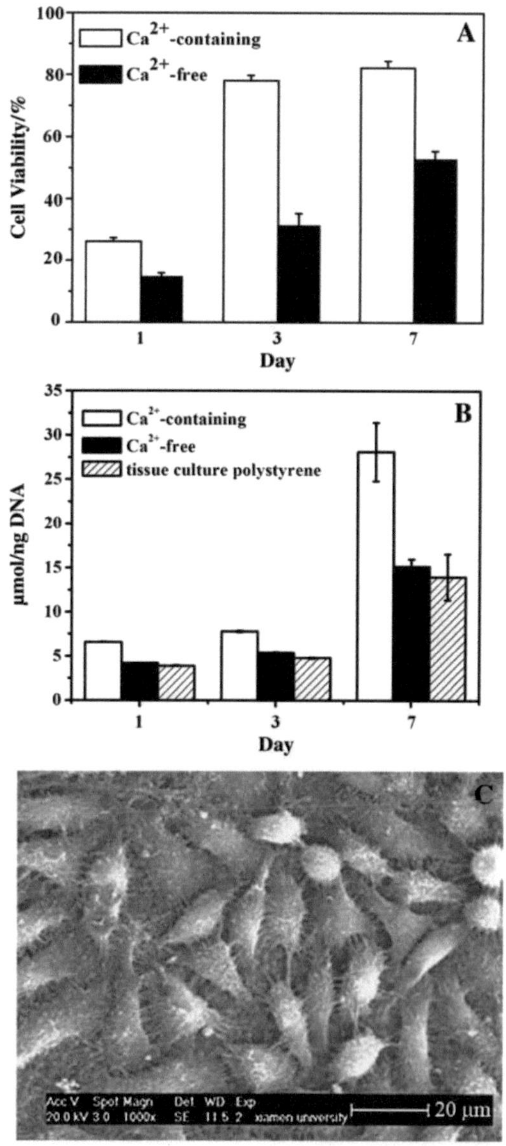

Figure 5. (A) MTT assay for proliferation of bone marrow-derived mesenchymal stem cells (BMSCs); (B) changes in ALP activity of BMSCs cultured on Ca^{2+}-containing gelatin/siloxane (GS) fiber mats and Ca^{2+}-free GS fiber mats at 1, 3, and 7 days, respectively. (C) SEM image of the Ca^{2+}-containing GS fiber mat seeded with BMSCs for 7 days [99].

membranes were promising candidates for the fabrication of contractile tissue engineered SMC-rich vascular medial layers [101] (**Figure 6**).

5.2. Drug delivery applications

Materials for drug delivery are often employed to release drug molecules into body tissues over extended periods of time. However, traditional drug delivery materials often do not

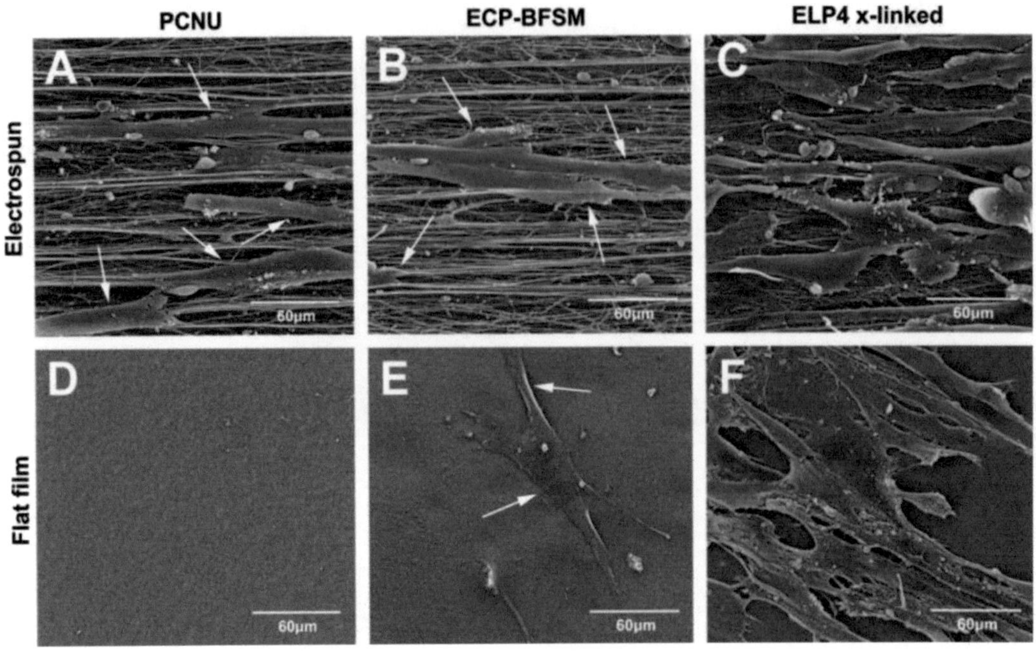

Figure 6. Scanning electron micrographs of SMC seeded materials after 1 week of culture (500× magnification). (A and D) polycarbonate-urethane (PCNU), (B and E) elastin cross-linking peptide bioactive fluorinated surface modifiers (ECP-BFSM) modified PCNU, and (C and F) elastin-like polypeptide-4 (ELP4) cross-linked aligned electrospun and flat film materials, respectively. Arrows indicate the edges of SMCs on the PCNU and ECP-BFSM modified materials [101].

control the rate of release of drugs effectively, while their poor drug absorption properties mean that they can only be used to deliver relatively low loadings of drug. Nanofibers are well suited as drug delivery vehicles, because their fine nanostructures are capable of absorbing significant amounts of drug molecules in a uniform manner. Gradual degradation of these nanofibers in the body then allows for gradual release of the drug in a controlled manner.

Polylactide-polyglycolide (PLGA) and collagen have been electrospun into sandwich structured drug-loaded membranes, with PLGA/collagen used for the surface layers, and PLGA/drugs contained in their core layer. The ability of these sandwich structured membranes to release vancomycin, gentamicin, and lidocaine in vitro was investigated. These results showed that these membranes could deliver therapeutic concentrations of vancomycin and gentamicin to human fibroblasts, ranging from 37 to 100% and 30 to 100% over 3 and 4 week periods, respectively [74] (**Figure 7**).

Meng et al. used electrospinning to uniformly distribute the drug Fenbufen throughout the structure of a nanofiber membrane prepared from gelatin and polylactide. The effect of gelatin concentration, fiber orientation, cross-linking time, and buffer pH on the sustained release of Fenbufen was successfully determined [102] (**Figure 8**). Huang et al. used proanthocyanidins as a cross-linking agent to prepare nanofibers as drug delivery vectors to deliver vitamin C magnesium phosphate. This study showed that the presence of the proanthocyanidin resulted in an increase in drug loading, which was important in maintaining a consistent rate of drug release. In addition, gelatin nanofibers that were cross-linked by proanthocyanidins were also seen to promote the proliferation of L929 cells. Therefore, this type of drug-loaded nanofibers material could potentially be used for controlled drug delivery, as well as for the promotion of wound healing [103] (**Figure 9**).

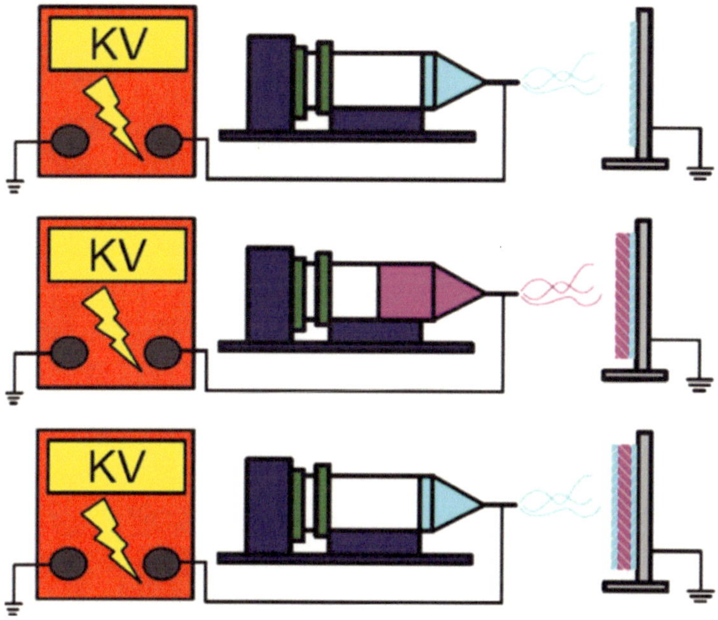

Figure 7. Apparatus used for electrospinning of sandwich structured membranes [74].

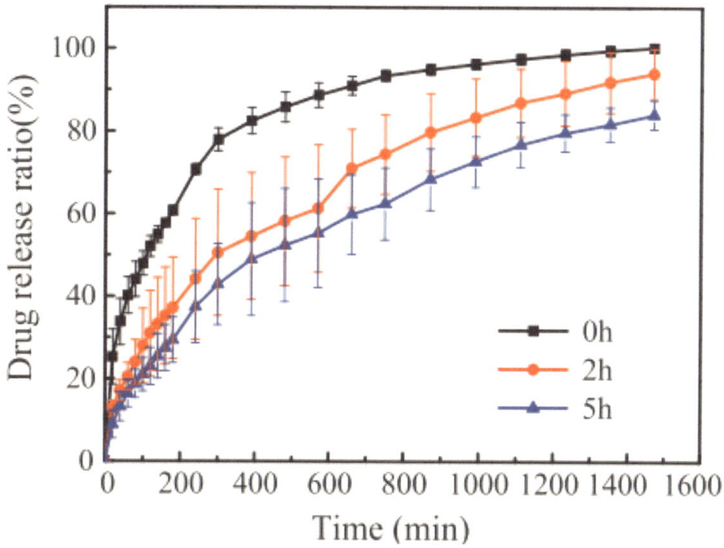

Figure 8. Release curves of FBF from PLGA/gelatin (9/1) electrospun nanofibers with different cross-linking times (0, 2, and 5 h) [102].

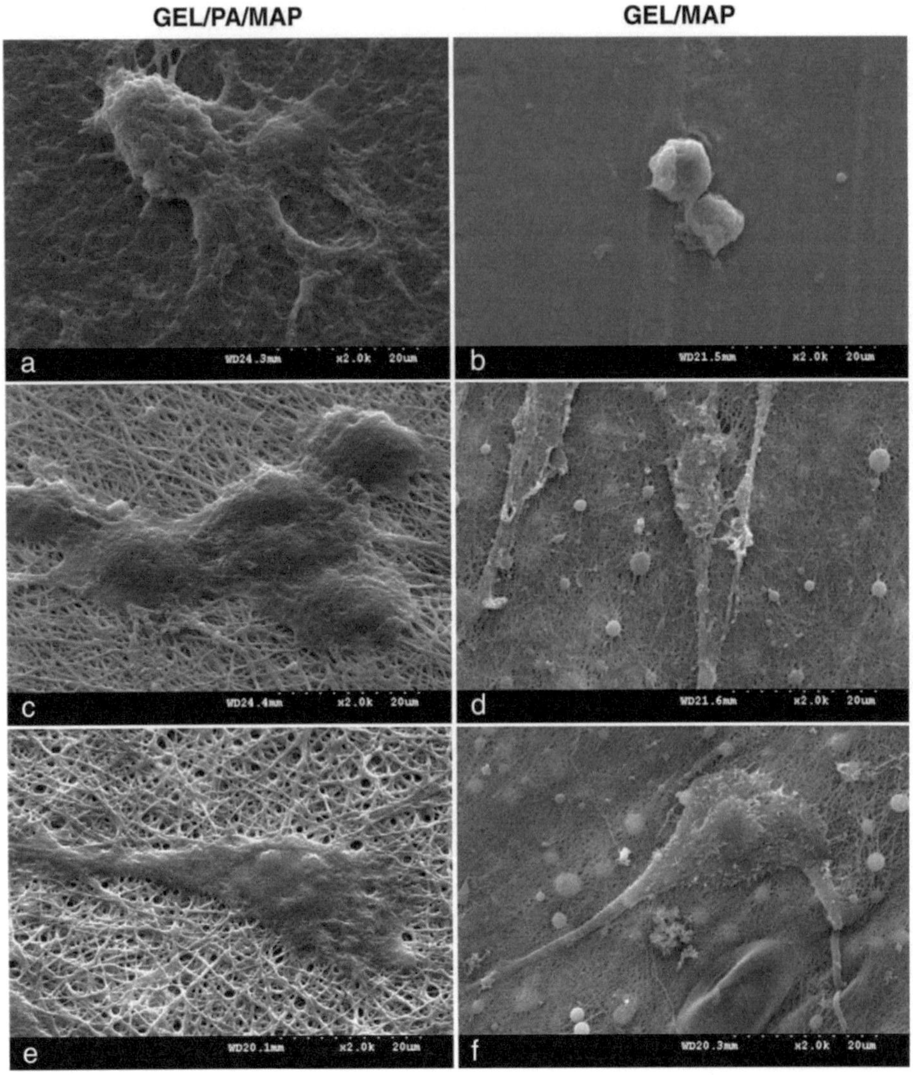

Figure 9. SEM photographs of the L929 cells cultured with gelatin (GEL)/magnesium ascorbyl phosphate (MAP) and GEL/proanthocyanidin (PA)/MAP membranes cross-linked by 50 wt.% glutaraldehyde vapor for 0, 15, and 45 min. (a) GEL/PA/MAP, 0 min. (b) GEL/MAP, 0 min. (c) GEL/PA/MAP, 15 min. (d) GEL/MAP, 15 min. (e) GEL/PA/MAP, 45 min. (f) GEL/MAP, 45 min [103].

Khadka et al. prepared electrospun composite nanofibers derived from a recombinant elastin-like peptide (ELP), or from a mixture of ELP with synthetic polypeptide and co-poly(L-glutamic acid$_4$, L-tyrosine$_1$) (PLEY). These materials contained numerous pores on their nanofibers surfaces, which ranged in diameter from <1 to 0.5 μm. Consequently, drugs were doped into the pores of these nanofibers materials and their use as potential drug delivery systems explored [104] (**Figure 10**).

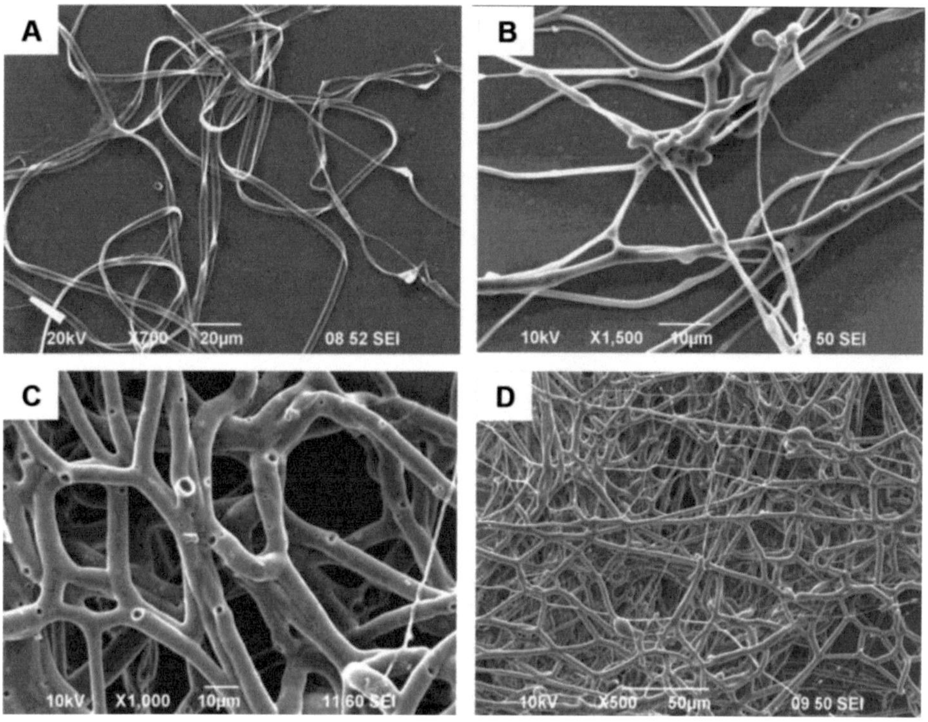

Figure 10. Scanning electron microscope image of a nonwoven fibers electrospun from a blended feedstock. In each case the nominal final polymer concentration was 48% (w/v) (A) (ELP)-C2:PLEY::2:1, 20 μm scale bar, (B) ELP-C2:PLEY::1:1, 10 μm scale bar, (C) ELP-C2:PLEY::2:3, 10 μm scale bar, (D) ELP-C2:PLEY::1:2, 50 μm scale bar. Note that the mean fiber diameter was ~5 μm in panels A, B and D, but ~10 μm in panel C [104].

5.3. Wound dressing applications

Collagen, gelatin, or polypeptide nanofiber membranes exhibit high porosities, small pore diameters, large surface areas, and fine microstructures that affords them good biocompatibility, biodegradability, biological adhesiveness, and moisture absorption properties. Therefore, these materials can be used to prepare dressings that keep wounds moist and which help prevent bacterial infection. Composite nanofiber materials prepared from gelatin and other materials such as fungicides, inflammatory drugs, and growth factors, can improve the performance of wound dressings. This is because they effectively improve the speed of wound hemostasis and healing by reducing its exposure to the external environment and protecting it from exposure to air.

Zine et al. prepared composite nanofibers by electrospinning a mixture of poly3-hydroxybutyric acid-co-3-hydroxyvaleric acid (PHBV), graphene oxide (GO), and collagen. GO was used to prepare these nanofibers to increase their mechanical strength and convey antibacterial activity against pathogenic bacteria such as *Escherichia coli* and *Staphylococcu aureus*. Collagen was included with the aim of enhancing cell proliferation, without affecting the composite materials mechanical strength or porosity. Subsequent biological results revealed that these nanofiber membranes generally had good wound healing properties [75] (**Figure 11**).

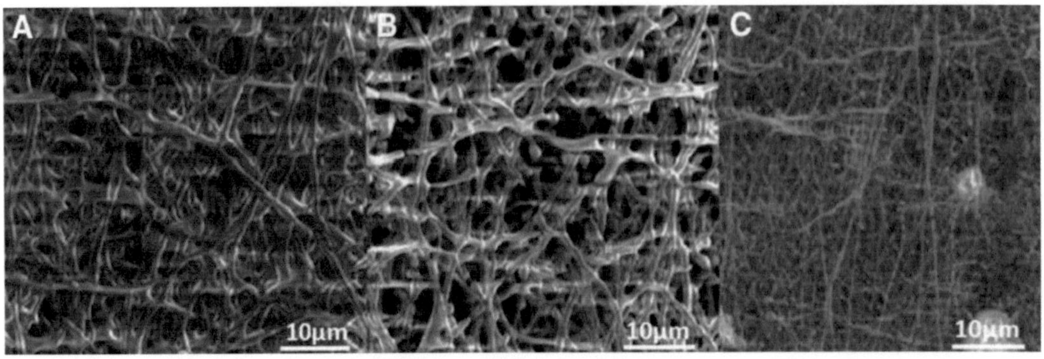

Figure 11. Scanning electron microscopy images of polymer (A) PHBV, (B) PHBV + GO, and (C) PHBV + GO + collagen [75].

Rujitanaroj et al. dissolved 22% (w/v) gelatin and 2.5 wt.% nitrate silver in 70 vol% of acetic acid to provide a stock solution for electrospinning nanofibers wound dressings with an average diameter of 280 nm. Glutaraldehyde was used as a cross-linking agent to improve the stability of the composite material in a moisture-rich environment. This nanofibers material was shown to have good sustained-release properties for Ag^{+}, resulting in these materials displaying good antimicrobial properties against *Pseudomonas aeruginosa* and *S. aureus* [105] (**Figure 12**).

Dubsky et al. used gelatin and polycaprolactone to prepare composite electrospun nanofibers, with subsequent cell culture experiments showing that these nanofibers can promoted cell adhesion and proliferation effectively. These composite nanofibers and medical gauze were applied to wounds of injured mice, with control experiments showing that inclusion of the nanofibers resulted in faster healing rates [106].

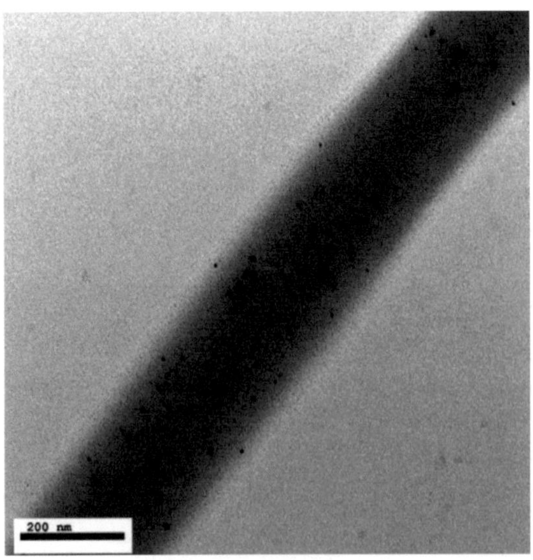

Figure 12. Selected TEM image of an electrospun fiber from the $AgNO_3$-containing gelatin solution that had been aged for 12 h [105].

5.4. Restorative materials for periodontal applications

The application of nanofiber membrane as periodontal restorative materials is an interdisciplinary field that spans the areas of tissue engineering and wound dressings. Periodontal repair requires membrane materials that prevent epithelial cells and connective tissue from growing into defect areas that can result in the creation of space for the co-migration and proliferation of periodontal ligament cells. Materials used for this dental application must be mechanically robust, exhibit good biodegradability profiles, and present robust three-dimensional nanostructures that are biocompatible with cell tissues. Therefore, collagen, gelatin, or polypeptide-derived nanofibers materials are potentially good choices for periodontal restorative applications.

Bottino et al. reported that a novel functionally graded membrane (FGM) could be prepared via sequential multilayer electrospinning. This FGM consisted of five different component layers, with each layer comprised of PLA:GEL+10 wt.% n-HAp, PLCL:PLA:GEL, pure PLCL, PLCL:PLA:GEL, and PLA:GEL+25 wt.% MET, respectively. Gelatin was used to enhance the bioactivity of this FGM, with poly-(DL-lactide-co-e-caprolactone) used to strengthen its

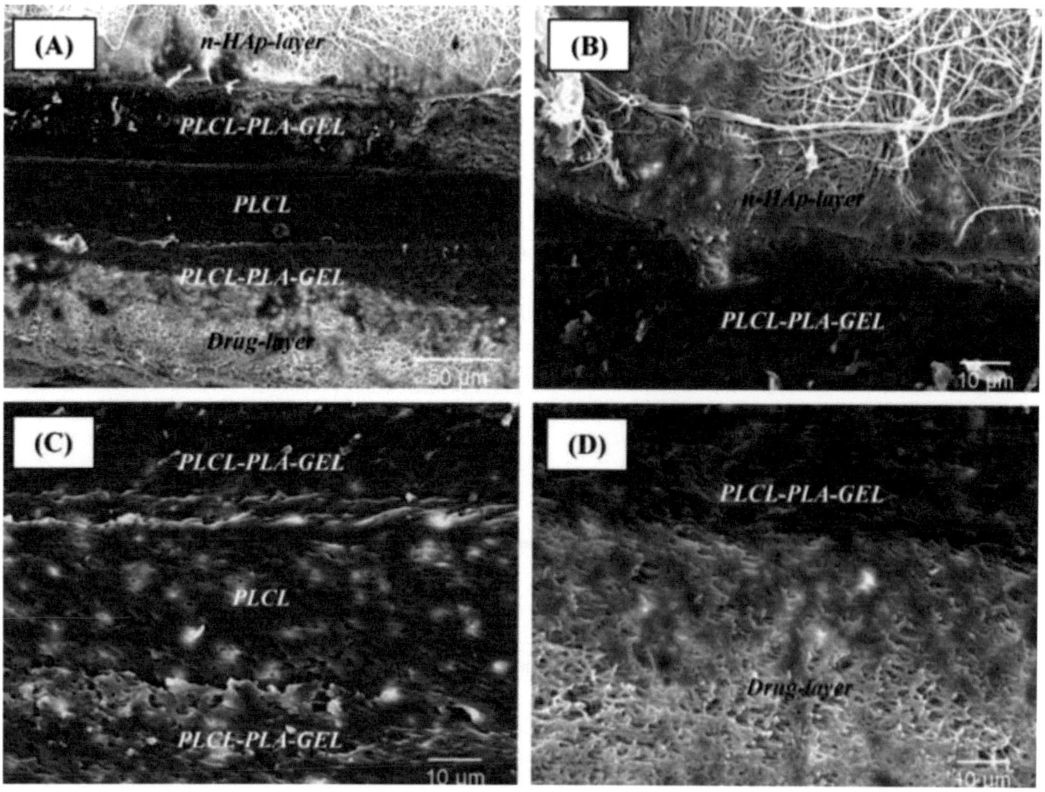

Figure 13. Cross-section SEM micrographs of the FGM processed via multilayering electrospinning. (A) General view of the FGM; (B) n-HAp-containing layer/PLCL:PLA:GEL interface; (C) CL structure; (D) MET-loaded layer/PLCL:PLA:GEL interface [107].

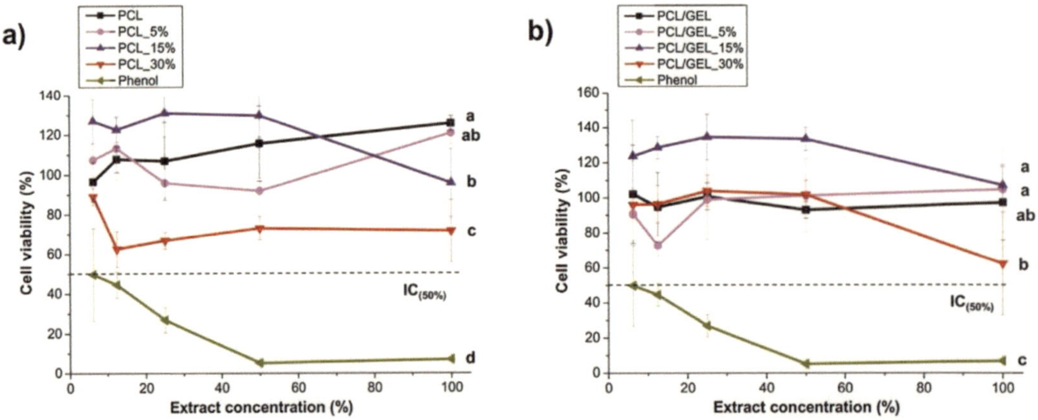

c)

Membranes	ZnO content (wt.%)			
	0	5	15	30
PCL	A 125.5 (±3.8) a	A 120.6 (±5.7) a	A 95.6 (±17.2) ab	A 71.5 (±15.7) b
PCL/GEL	B 97.3 (±21.4) ab	A 104.7 (±24.1) a	A 107.1 (±10.4) a	A 62.3 (±29.5) b

Figure 14. Cytotoxicity assays results on PCL (a) and PCL/GEL-based (b) membranes and cell viability [means (%) and standard deviations (±SD)] after exposure to concentrated (100%) extracts (c) [108].

mechanical properties and metronidazole benzoate (MET) included to prevent bacterial infection. n-HAp was incorporated into the PLA:GEL membrane to mimic the collagen-HAp matrix that was present in bone and enhance the composites osteoconductive behavior. This formulation afforded an electrospun FGM with excellent mechanical integrity, biodegradability, and good cell-membrane interactions that was explored as a periodontal restorative material [107] (**Figure 13**).

Mixtures of polycaprolactone and gelatin and ZnO in hexafluoropropanol had also been electrospun into nanofibers to afford ZnO-loaded electrospun membranes that possessed good biocompatibility, stretching ability, antibacterial activity, as potentially useful materials for periodontal regeneration [108] (**Figure 14**).

6. Outlook

Collagen and gelatin (and their derived polypeptides) are natural biopolymers that exhibit good biocompatibility, biodegradability, and low immunogenicity, as well as being excellent materials as hosts for cell and tissue growth. With the development of collagen in medical application, many processing methods are required to ensure that their application can be fully realized. The rapid evolution of electrospinning techniques is well suited to meet this need, enabling nanofiber membranes with three-dimensional pore structures, which imitated the microstructure of the extracellular matrix. Consequently, these electrospinning techniques have attracted increasing attention for applications in the fields of tissue engineering, drug delivery, wound dressing, nerve regeneration, periodontal regeneration, and vascular reconstruction. The organic solvent used in electrospinning processes has potential toxicity issues, so the development of approaches that allow electrospinning to be carried out under aqueous conditions is highly desirable. Collagen, gelatin, or polypeptide nanofibers have been used to prepare biopolymers for applications in many different biomedical fields, including wound repair, artificial skin, and drug delivery. In order to satisfy a large number of medical applications, it will be necessary to develop more efficient electrospinning technics that produce large amounts of material with optimal microscopic structures. Current problems need to be resolved, include low efficiency, low mechanical strength, and poor spinning reproducibility. In addition, a lot of effort should be done in spinneret design, collecting device optimization, and solution delivery techniques. If these issues can be overcome, it is anticipated that collagen-derived nanofiber materials will have a major role for biomedical and biotechnological applications.

Author details

Wei Peng Lu[1] and Yanchuan Guo[1,2]*

*Address all correspondence to: yanchuanguo@mail.ipc.ac.cn

1 Key Laboratory of Photochemical Conversion and Optoelectronic Materials, Technical Institute of Physics and Chemistry, Chinese Academy of Science, Beijing, China

2 University of Chinese Academy of Sciences, Beijing, PR China

References

[1] Chen Z, Mo X, Qing F. Electrospinning of collagen-chitosan complex. Materials Letters. 2007;**61**(16):3490-3494

[2] Chakrapani VY, Gnanamani A, Giridev VR, Madhusoothanan M, Sekaran G. Electrospinning of type I collagen and PCL nanofibers using acetic acid. Journal of Applied Polymer Science. 2012;**125**(4):3221-3227

[3] Zhang M, Wang J, Xu W, Luan J, Li X, Zhang Y, Dong H, Sun D. The mechanical property of *Rana chensinensis* skin collagen/poly(L-lactide) fibers membrane. Materials Letters. 2015;**139**:467-470

[4] Homaeigohar SS, Mahdavi H, Elbahri M. Extraordinarily water permeable sol-gel formed nanocomposite nanofibers membranes. Journal of Colloid and Interface Science. 2012;**366**(1):51-56

[5] Chang B, Guan D, Tian Y, Yang Z, Dong X. Convenient synthesis of porous carbon nanospheres with tunable pore structure and excellent adsorption capacity. Journal of Hazardous Materials. 2013;**262**:256-264

[6] Liang D, Hsiao BS, Chu B. Functional electrospun nanofibers scaffolds for biomedical applications. Advanced Drug Delivery Reviews. 2007;**59**(14):1392-1412

[7] Maheshwari S, Chang H-C. Assembly of multi-stranded nanofiber threads through AC electrospinning. Advanced Materials. 2009;**21**(3):349-354

[8] Forward KM, Flores A, Rutledge GC. Production of core/shell fibers by electrospinning from a free surface. Chemical Engineering Science. 2013;**104**(0):250-259

[9] Zhou T, Wang N, Xue Y, Ding T, Liu X, Mo X, Sun J. Electrospun tilapia collagen nanofibers accelerating wound healing via inducing keratinocytes proliferation and differentiation. Colloids and Surfaces B: Biointerfaces. 2016;**143**:415-422

[10] Jing X, Mi H-Y, Cordie TM, Salick MR, Peng X-F, Turng L-S. Fabrication of shish–kebab structured poly(ε-caprolactone) electrospun nanofibers that mimic collagen fibrils: Effect of solvents and matrigel functionalization. Polymer. 2014;**55**(21):5396-5406

[11] Hiraoka K. Fundamentals of electrospray. In: Hiraoka K, editor. Fundamentals of Mass Spectrometry. New York, NY: Springer New York; 2013. pp. 145-171

[12] Kong CS, Lee TH, Lee SH, Kim HS. Nano-web formation by the electrospinning at various electric fields. Journal of Materials Science. 2007;**42**(19):8106-8112

[13] Goh Y-F, Shakir I, Hussain R. Electrospun fibers for tissue engineering, drug delivery, and wound dressing. Journal of Materials Science. 2013;**48**(8):3027-3054

[14] Mohammadzadehmoghadam S, Dong Y, Jeffery Davies I. Recent progress in electrospun nanofibers: Reinforcement effect and mechanical performance. Journal of Polymer Science Part B: Polymer Physics. 2015;**53**(17):1171-1212

[15] Cipitria A, Skelton A, Dargaville TR, Dalton PD, Hutmacher DW. Design, fabrication and characterization of PCL electrospun scaffolds-a review. Journal of Materials Chemistry. 2011;**21**(26):9419-9453

[16] Choi DJ, Choi SM, Kang HY, Min H-J, Lee R, Ikram M, Subhan F, Jin SW, Jeong YH, Kwak J-Y, Yoon S. Bioactive fish collagen/polycaprolactone composite nanofibers scaffolds fabricated by electrospinning for 3D cell culture. Journal of Biotechnology. 2015;**205**:47-58

[17] Arafat MT, Tronci G, Yin J, Wood DJ, Russell SJ. Biomimetic wet-stable fibres via wet spinning and diacid-based crosslinking of collagen triple helices. Polymer. 2015;**77**:102-112

[18] Law JX, Musa F, Ruszymah BHI, El Haj AJ, Yang Y. A comparative study of skin cell activities in collagen and fibrin constructs. Medical Engineering & Physics. 2016;**38**(9):854-861

[19] Gelse K, Pöschl E, Aigner T. Collagens—Structure, function, and biosynthesis. Advanced Drug Delivery Reviews. 2003;**55**(12):1531-1546

[20] Shoulders MD, Raines RT. Collagen structure and stability. Annual Review of Biochemistry. 2009;**78**(1):929-958

[21] Arslan YE, Sezgin Arslan T, Derkus B, Emregul E, Emregul KC. Fabrication of human hair keratin/jellyfish collagen/eggshell-derived hydroxyapatite osteoinductive biocomposite scaffolds for bone tissue engineering: From waste to regenerative medicine products. Colloids and Surfaces B: Biointerfaces. 2017;**154**:160-170

[22] Wei S, David CM, Sunxi W, Tong S, Guangzhao M, Weiping R. Electrospun polyvinyl alcohol-collagen-hydroxyapatite nanofibers: A biomimetic extracellular matrix for osteoblastic cells. Nanotechnology. 2012;**23**(11):115101

[23] Zulkifli FH, Jahir Hussain FS, Abdull Rasad MSB, Mohd Yusoff M. In vitro degradation study of novel HEC/PVA/collagen nanofibers scaffold for skin tissue engineering applications. Polymer Degradation and Stability. 2014;**110**:473-481

[24] Kuttappan S, Mathew D, Nair MB. Biomimetic composite scaffolds containing bioceramics and collagen/gelatin for bone tissue engineering—A mini review. International Journal of Biological Macromolecules. 2016;**93**:1390-1401

[25] Johnson BJ. Synthesis, structure, and biological properties of sequential polypeptides. Journal of Pharmaceutical Sciences. 1974;**63**(3):313-327

[26] Fontaine-Vive F, Merzel F, Johnson MR, Kearley GJ. Collagen and component polypeptides: Low frequency and amide vibrations. Chemical Physics. 2009;**355**(2):141-148

[27] Tabata Y, Ikada Y. Protein release from gelatin matrices. Advanced Drug Delivery Reviews. 1998;**31**(3):287-301

[28] Wang Y, Zhang W, Yuan J, Shen J. Differences in cytocompatibility between collagen, gelatin and keratin. Materials Science and Engineering: C. 2016;**59**:30-34

[29] Nieto-Suárez M, López-Quintela MA, Lazzari M. Preparation and characterization of crosslinked chitosan/gelatin scaffolds by ice segregation induced self-assembly. Carbohydrate Polymers. 2016;**141**:175-183

[30] Huang X, Zhang Y, Zhang X, Xu L, Chen X, Wei S. Influence of radiation crosslinked carboxymethyl-chitosan/gelatin hydrogel on cutaneous wound healing. Materials Science and Engineering: C. 2013;**33**(8):4816-4824

[31] Sarika PR, Cinthya K, Jayakrishnan A, Anilkumar PR, James NR. Modified gum arabic cross-linked gelatin scaffold for biomedical applications. Materials Science and Engineering: C. 2014;**43**:272-279

[32] Hou H, Li B, Zhang Z, Xue C, Yu G, Wang J, Bao Y, Bu L, Sun J, Peng Z, Su S. Moisture absorption and retention properties, and activity in alleviating skin photodamage of collagen polypeptide from marine fish skin. Food Chemistry. 2012;**135**(3):1432-1439

[33] Chen T, Hou H. Protective effect of gelatin polypeptides from Pacific cod (*Gadus macrocephalus*) against UV irradiation-induced damages by inhibiting inflammation and improving transforming growth factor-β/Smad signaling pathway. Journal of Photochemistry and Photobiology B: Biology. 2016;**162**:633-640

[34] Rizvi MS, Kumar P, Katti DS, Pal A. Mathematical model of mechanical behavior of micro/nanofibers materials designed for extracellular matrix substitutes. Acta Biomaterialia. 2012;**8**(11):4111-4122

[35] Hong Y, Huber A, Takanari K, Amoroso NJ, Hashizume R, Badylak SF, Wagner WR. Mechanical properties and in vivo behavior of a biodegradable synthetic polymer microfiber–extracellular matrix hydrogel biohybrid scaffold. Biomaterials. 2011;**32**(13): 3387-3394

[36] Arnal-Pastor M, Ramos CM, Garnés MP, Pradas MM, Lluch AV. Corrigendum to electrospun adherent-antiadherent bilayered membranes based on crosslinked hyaluronic acid for advanced tissue engineering applications. Materials Science and Engineering: C. 2014;**37**:405

[37] Arnal-Pastor M, Martínez Ramos C, Pérez Garnés M, Monleón Pradas M, Vallés Lluch A. Electrospun adherent–antiadherent bilayered membranes based on cross-linked hyaluronic acid for advanced tissue engineering applications. Materials Science and Engineering: C. 2013;**33**(7):4086-4093

[38] Soukup K, Petráš D, Topka P, Slobodian P, Šolcová O. Preparation and characterization of electrospun poly(p-phenylene oxide) membranes. Catalysis Today. 2012;**193**(1):165-171

[39] Kostakova E, Meszaros L, Gregr J. Composite nanofibers produced by modified needleless electrospinning. Materials Letters. 2009;**63**(28):2419-2422

[40] Vysloužilová L, Buzgo M, Pokorný P, Chvojka J, Míčková A, Rampichová M, Kula J, Pejchar K, Bílek M, Lukáš D, Amler E. Needleless coaxial electrospinning: A novel approach to mass production of coaxial nanofibers. International Journal of Pharmaceutics. 2017;**516**(1):293-300

[41] Hapidin DA, Saleh I, Munir MM, Khairurrijal. Design and development of a series-configuration mazzilli zero voltage switching flyback converter as a high-voltage power supply for needleless electrospinning. Procedia Engineering. 2017;**170**:509-515

[42] Li D, Wu T, He N, Wang J, Chen W, He L, Huang C, Ei-Hamshary HA, Al-Deyab SS, Ke Q, Mo X. Three-dimensional polycaprolactone scaffold via needleless electrospinning promotes cell proliferation and infiltration. Colloids and Surfaces B: Biointerfaces. 2014;**121**:432-443

[43] He JH, Liu Y, Xu L, Yu JY, Sun G. BioMimic fabrication of electrospun nanofibers with high-throughput. Chaos, Solitons & Fractals. 2008;**37**(3):643-651

[44] Huang ZM, Zhang YZ, Kotaki M, Ramakrishna S. A review on polymer nanofibers by electrospinning and their applications in nanocomposites. Composites Science and Technology. 2003;**63**(15):2223-2253

[45] Khamforoush M, Mahjob M. Modification of the rotating jet method to generate highly aligned electrospun nanofibers. Materials Letters. 2011;**65**(3):453-455

[46] Yan H, Liu L, Zhang Z. Continually fabricating staple yarns with aligned electrospun polyacrylonitrile nanofibers. Materials Letters. 2011;**65**(15):2419-2421

[47] Matthews JA, Wnek GE, Simpson DG, Bowlin GL. Electrospinning of collagen nanofibers. Biomacromolecules. 2002;**3**(2):232-238

[48] Chen R, Li XQ, Ke QF, He CL, Mo XM. Fabrication and characterization of collagen (shell) /thermoplastic polyurethane (core) composite nanofibers by coaxial electrospinning. e-Polymers. 2010;**10**(1):67-74

[49] Chen R, Huang C, Ke QF, He CL, Wang HS, Mo XM. Preparation and characterization of coaxial electrospun thermoplastic polyurethane/collagen compound nanofibers for tissue engineering applications. Colloids and Surfaces B: Biointerfaces. 2010;**79**(2):315-325

[50] Huang ZM, Zhang YZ, Ramakrishna S, Lim CT. Electrospinning and mechanical characterization of gelatin nanofibers. Polymer. 2004;**45**(15):5361-5368

[51] Su Y, Li XQ, Liu SP, Wang HS, He CL. Fabrication and properties of PLLA-gelatin nanofibers by electrospinning. Journal of Applied Polymer Science. 2010;**117**(1):542-547

[52] Li MY, Mondrinos MJ, Gandhi MR, Ko FK, Weiss AS, Lelkes PI. Electrospun protein fibers as matrices for tissue engineering. Biomaterials. 2005;**26**(30):5999-6008

[53] Wnek GE, Carr ME, Simpson DG, Bowlin GL. Electrospinning of nanofiber fibrinogen structures. Nano Letters. 2007;**3**(2):213-216

[54] Gugutkov D, Gustavsson J, Ginebra MP, Altankov G. Fibrinogen nanofibers for guiding endothelial cell behavior. Biomaterials. 2013;**1**(10):1065-1073

[55] Gunn J, Zhang MQ. Polyblend nanofibers for biomedical applications: Perspectives and challenges. Trends in Biotechnology. 2010;**28**(4):189-197

[56] He C, Nie W, Feng W. Engineering of biomimetic nanofibers matrices for drug delivery and tissue engineering. Journal of Materials Chemistry B. 2014;**2**(45):7828-7848

[57] Vohra V, Calzaferri G, Destri S, Pasini M, Porzio W, Botta C. Toward white light emission through efficient two-step energy transfer in hybrid nanofibers. ACS Nano. 2010;**4**(3):1409-1416

[58] Chen JP, Chang GY, Chen JK. Electrospun collagen/chitosan nanofibers membrane as wound dressing. Colloids and Surfaces A: Physicochemical and Engineering Aspects. 2008;**313**:183-188

[59] Gu SY, Wang ZM, Ren J, Zhang C-Y. Electrospinning of gelatin and gelatin/poly(L-lactide) blend and its characteristics for wound dressing. Materials Science and Engineering: C. 2009;**29**(6):1822-1828

[60] Chen Z, Cao L, Wang L, Zhu H, Jiang H. Effect of fiber structure on the properties of the electrospun hybrid membranes composed of poly(ε-caprolactone) and gelatin. Journal of Applied Polymer Science. 2013;**127**(6):4225-4232

[61] Kazanci M. Solvent and temperature effects on folding of electrospun collagen nanofibers. Materials Letters. 2014;**130**:223-226

[62] Dulnik J, Denis P, Sajkiewicz P, Kołbuk D, Choińska E. Biodegradation of bicomponent PCL/gelatin and PCL/collagen nanofibers electrospun from alternative solvent system. Polymer Degradation and Stability. 2016;**130**:10-21

[63] Lu W, Xu H, Zhang B, Ma M, Guo Y. The preparation of chitosan oligosaccharide/alginate sodium/gelatin nanofibers by spiral-electrospinning. Journal of Nanoscience and Nanotechnology. 2015;**15**:1-5

[64] Wang X, Ding B, Yu J, Yang J. Large-scale fabrication of two-dimensional spider-web-like gelatin nano-nets via electro-netting. Colloids and Surfaces B: Biointerfaces. 2011;**86**(2):345-352

[65] An K, Liu H, Guo S, Kumar DNT, Wang Q. Preparation of fish gelatin and fish gelatin/poly(L-lactide) nanofibers by electrospinning. International Journal of Biological Macromolecules. 2010;**47**(3):380-388

[66] Huang L, McMillan RA, Apkarian RP, Pourdeyhimi B, Conticello VP, Chaikof EL. Generation of synthetic elastin-mimetic small diameter fibers and fiber networks. Macromolecules. 2000;**33**(8):2989-2997

[67] Jose MV, Thomas V, Dean DR, Nyairo E. Fabrication and characterization of aligned nanofibers PLGA/collagen blends as bone tissue scaffolds. Polymer. 2009;**50**(15):3778-3785

[68] Meng W, Kim SY, Yuan J, Kim JC, Kwon OH, Kawazoe N, Chen GP, Ito Y, Kang IK. Electrospun PHBV/collagen composite nanofibers scaffolds for tissue engineering. Journal of Biomaterials Science Polymer Edition. 2007;**18**(1):81

[69] Venugopal J, Zhang YZ, Ramakrishna S. Fabrication of modified and functionalized polycaprolactone nanofibre scaffolds for vascular tissue engineering. Nanotechnology. 2005;**16**(10):2138-2142

[70] Chiu JB, Liu C, Hsiao BS, Chu B, Hadjiargyrou M. Functionalization of poly(L-lactide) nanofibers scaffolds with bioactive collagen molecules. Journal of Biomedical Materials Research Part A. 2010;**83A**(4):1117-1127

[71] Yu CC, Chang JJ, Lee YH, Lin YC, Wu MH, Yang MC, Chien CT. Electrospun scaffolds composing of alginate, chitosan, collagen and hydroxyapatite for applying in bone tissue engineering. Materials Letters. 2013;**93**:133-136

[72] Choi JS, Lee SJ, Christ GJ, Atala A, Yoo JJ. The influence of electrospun aligned poly(ε-caprolactone)/collagen nanofiber meshes on the formation of self-aligned skeletal muscle myotubes. Biomaterials. 2008;**29**(19):2899-2906

[73] Lee BK, Ju YM, Cho JG, Jackson JD, Lee SJ, Atala A, Yoo JJ. End-to-side neurorrhaphy using an electrospun PCL/collagen nerve conduit for complex peripheral motor nerve regeneration. Biomaterials. 2012;**33**(35):9027-9036

[74] Chen DW, Hsu Y-H, Liao J-Y, Liu S-J, Chen J-K, Ueng SW-N. Sustainable release of vancomycin, gentamicin and lidocaine from novel electrospun sandwich-structured PLGA/collagen nanofibers membranes. International Journal of Pharmaceutics. 2012; **430**(1):335-341

[75] Zine R, Sinha M. Nanofibers poly (3-hydroxybutyrate-co-3-hydroxyvaletate)/collagen/ grapheme oxide scaffolds for wound coverage. Materials Science and Engineering: C. 2017;**80**:129-134

[76] Chen ZG, Wang PW, Wei B, Mo XM, Cui FZ. Electrospun collagen-chitosan nanofiber: A biomimetic extracellular matrix for endothelial cell and smooth muscle cell. Acta Biomaterialia. 2010;**6**(2):372-382

[77] Dong B, Arnoult O, Smith ME, Wnek GE. Electrospinning of collagen nanofiber scaffolds from benign solvents. Macromolecular Rapid Communications. 2009;**30**(7):539-542

[78] Heydarkhan-Hagvall S, Schenke-Layland K, Dhanasopon AP, Rofail F, Smith H, Wu BM, Shemin R, Beygui RE, MacLellan WR. Three-dimensional electrospun ECM-based hybrid scaffolds for cardiovascular tissue engineering. Biomaterials. 2008;**29**(19):2907-2914

[79] Lee SJ, Liu J, Oh SH, Soker S, Atala A, Yoo JJ. Development of a composite vascular scaffolding system that withstands physiological vascular conditions. Biomaterials. 2008;**29**(19):2891-2898

[80] Powell HM, Boyce ST. Engineered human skin fabricated using electrospun collagen-PCL blends: Morphogenesis and mechanical properties. Tissue Engineering. Part A. 2009;**15**(8):2177-2187

[81] Huang L, Nagapudi K, Apkarian RP, Chaikof EL. Engineered collagen-PEO nanofibers and fabrics. Journal of Biomaterials Science Polymer Edition. 2001;**12**(9):979-993

[82] Schnell E, Klinkhammer K, Balzer S, Brook G, Klee D, Dalton P, Mey J. Guidance of glial cell migration and axonal growth on electrospun nanofibers of poly-epsilon-caprolactone and a collagen/poly-epsilon-caprolactone blend. Biomaterials. 2007;**28**(19):3012-3025

[83] He W, Yong T, Teo WE, Ma ZW, Ramakrishna S. Fabrication and endothelialization of collagen-blended biodegradable polymer nanofibers: Potential vascular graft for blood vessel tissue engineering. Tissue Engineering Part A. 2005;**11**(10):1574-1588

[84] Kwon IK, Matsuda T. Co-electrospun nanofiber fabrics of poly(L-lactide-co-epsilon-caprolactone) with type I collagen or heparin. Biomacromolecules. 2005;**6**(4):2096-2105

[85] Stitzel J, Liu L, Lee SJ, Komura M, Berry J, Soker S, Lim G, Van Dyke M, Czerw R, Yoo JJ, Atala A. Controlled fabrication of a biological vascular substitute. Biomaterials. 2006; **27**(7):1088-1094

[86] Lee SJ, Yoo JJ, Lim GJ, Atala A, Stitze J. In vitro evaluation of electrospun nanofiber scaffolds for vascular graft application. Journal of Biomedical Materials Research Part A. 2007;**83**(4):999-1008

[87] Jeong SI, Lee AY, Lee YM, Shin H. Electrospun gelatin/poly(L-lactide-co-epsilon-caprolactone) nanofibers for mechanically functional tissue-engineering scaffolds. Journal of Biomaterials Science Polymer Edition. 2008;**19**(3):339-357

[88] Zhang YZ, Ouyang HW, Lim CT, Ramakrishna S, Huang ZM. Electrospinning of gelatin fibers and gelatin/PCL composite fibers scaffolds. Journal of Biomedical Materials Research Part B Applied Biomaterials. 2005;**72B**(1):156-165

[89] Zhao PC, Jiang HL, Pan H, Zhu KJ, Chen W. Biodegradable fibers scaffolds composed of gelatin coated poly(epsilon-caprolactone) prepared by coaxial electrospinning. Journal of Biomedical Materials Research Part A. 2007;**83**(2):372-382

[90] Gupta D, Venugopal J, Prabhakaran MP, Dev VRG, Low S, Choon AT, Ramakrishna S. Aligned and random nanofibers substrate for the in vitro culture of Schwann cells for neural tissue engineering. Acta Biomaterialia. 2009;**5**(7):2560-2569

[91] Deng LL, Zhang X, Li Y, Que F, Kang XF, Liu YY, Feng FQ, Zhang H. Characterization of gelatin/zein nanofibers by hybrid electrospinning. Food Hydrocolloids. 2018;**75**:72-80

[92] Tavassoli-Kafrani E, Goli SAH, Fathi M. Fabrication and characterization of electrospun gelatin nanofibers crosslinked with oxidized phenolic compounds. International Journal of Biological Macromolecules. 2017;**103**:1062-1068

[93] Jalaja K, Sreehari VS, Kumar PRA, Nirmala RJ. Graphene oxide decorated electrospun gelatin nanofibers: Fabrication, properties and applications. Materials Science & Engineering C Materials for Biological Applications. 2016;**64**:11-19

[94] Shi R, Geng H, Gong M, Ye JJ, Wu CG, Hu XH, Zhang LQ. Long-acting and broad-spectrum antimicrobial electrospun poly (ε-caprolactone)/gelatin micro/nanofibers for wound dressing. Journal of Colloid & Interface Science. 2017;**509**:275-284

[95] Talebian A, Mansourian A. Release of Vancomycin from electrospun gelatin/chitosan nanofibers. Materials Today Proceedings. 2017;**4**(7):7065-7069

[96] KarbalaeiMandi A, Shahrousvand M, Javadi HR, Ghollasi M, Norouz F, Kamali M, Salimi A. Neural differentiation of human induced pluripotent stem cells on polycaprolactone/gelatin bi-electrospun nanofibers. Materials Science & Engineering C. 2017;**78**:1195-1202

[97] Rather HA, Thakore R, Singh R, Jhala D, Singh S, Vasita R. Antioxidative study of cerium oxide nanoparticle functionalised PCL-Gelatin electrospun fibers for wound healing application. Bioactive Materials. 2017:1-11

[98] Ji W, Yang F, Ma J, Bouma MJ, Boerman OC, Chen Z, van den Beucken JJJP, Jansen JA. Incorporation of stromal cell-derived factor-1α in PCL/gelatin electrospun membranes for guided bone regeneration. Biomaterials. 2013;**34**(3):735-745

[99] Ren L, Wang J, Yang FY, Wang L, Wang D, Wang TX, Tian MM. Fabrication of gelatin–siloxane fibers mats via sol–gel and electrospinning procedure and its application for bone tissue engineering. Materials Science and Engineering: C. 2010;**30**(3):437-444

[100] Li M, Guo Y, Wei Y, MacDiarmid AG, Lelkes PI. Electrospinning polyaniline-contained gelatin nanofibers for tissue engineering applications. Biomaterials. 2006; **27**(13):2705-2715

[101] Blit PH, Battiston KG, Yang M, Paul Santerre J, Woodhouse KA. Electrospun elastin-like polypeptide enriched polyurethanes and their interactions with vascular smooth muscle cells. Acta Biomaterialia. 2012;**8**(7):2493-2503

[102] Meng ZX, Xu XX, Zheng W, Zhou HM, Li L, Zheng YF, Lou X. Preparation and characterization of electrospun PLGA/gelatin nanofibers as a potential drug delivery system. Colloids and Surfaces B: Biointerfaces. 2011;**84**(1):97-102

[103] Huang CH, Chi CY, Chen YS, Chen KY, Chen PL, Yao CH. Evaluation of proanthocyanidin-crosslinked electrospun gelatin nanofibers for drug delivering system. Materials Science and Engineering: C. 2012;**32**(8):2476-2483

[104] Khadka DB, Niesen MI, Devkota J, Koria P, Haynie DT. Unique electrospun fiber properties obtained by blending elastin-like peptides and highly-ionized peptides. Polymer. 2014;**55**(9):2163-2169

[105] Rujitanaroj P-O, Pimpha N, Supaphol P. Wound-dressing materials with antibacterial activity from electrospun gelatin fiber mats containing silver nanoparticles. Polymer. 2008;**49**(21):4723-4732

[106] Dubský M, Kubinová Š, Širc J, Voska L, Zajíček R, Zajícová A, Lesný P, Jirkovská A, Michálek J, Munzarová M, Holáň V, Syková E. Nanofibers prepared by needleless electrospinning technology as scaffolds for wound healing. Journal of Materials Science: Materials in Medicine. 2012;**23**(4):931-941

[107] Bottino MC, Thomas V, Janowski GM. A novel spatially designed and functionally graded electrospun membrane for periodontal regeneration. Acta Biomaterialia. 2011; **7**(1):216-224

[108] Münchow EA, Albuquerque MTP, Zero B, Kamocki K, Piva E, Gregory RL, Bottino MC. Development and characterization of novel ZnO-loaded electrospun membranes for periodontal regeneration. Dental Materials. 2015;**31**(9):1038-1051

Section 2

Nanofiber Properties and Applications

Chapter 4

Photochromic Nanofibers

Emriye Perrin Akçakoca Kumbasar,
Seniha Morsunbul and Simge Alır

Additional information is available at the end of the chapter

http://dx.doi.org/10.5772/intechopen.74663

Abstract

Photochromic compounds exhibit a reversible color change via UV irradiation. The use of photochromic nanofibers in the field of functional materials such as optical sensors, processing media, optical data storage devices, and functional components for smart surfaces can be attractive. This review chapter gives an overview of the production of photochromic nanofibers by electrospinning.

Keywords: nanofiber, photochromism, photochromic textile, electrospun, UV-responsive nanofibers

1. Introduction

Nanotechnology applications and nanofiber production have become increasingly important in terms of functional textile material production. Nanofiber surfaces have potential to be used in a wide range of applications, such as filtration, sensor, composite materials, medical textiles, etc., because of their wide surface area, high porosity, and possibilities of using different polymer types [1–3].

Nanofiber production can be achieved by different methods such as electrospinning, drawing, self-assembly, phase separation, and template synthesis. While the applications of all these methods are generally laboratory scale, the electrospinning process can be scaled, and it has a potential for industry processing [4]. Electrospinning method has also some advantages such as ease of application, low cost, and possibility of using many different polymers. Because of all these properties, electrospinning is perhaps the most preferred method for the nanofiber production [1, 3–7].

IntechOpen

© 2018 The Author(s). Licensee IntechOpen. This chapter is distributed under the terms of the Creative Commons Attribution License (http://creativecommons.org/licenses/by/3.0), which permits unrestricted use, distribution, and reproduction in any medium, provided the original work is properly cited.

Nanofiber surfaces may gain functional properties with the addition of different substances such as drugs, cosmetics, dyes, etc. in the polymer matrix, thereby increasing the industrial application possibilities [1, 5, 7–11]. A photochromic compound, which can change its color upon irradiation with ultraviolet (UV), is also one of the materials used in the nanofiber surface functionalization. This chapter describes the functionalization of nanofiber surfaces produced by electrospinning method with photochromic compounds.

2. Electrospinning

In the electrospinning process, the polymer solution or polymer melt is subjected to a high potential tension, and the polymers are electrically charged. The polymer solution jet, which comes from a fine nozzle, travels toward the target, which is loaded with the reverse polarity and positioned opposite the nozzle. During this flow, the solvent evaporates (or begins to solidify if electrospinning from the melt is applied), and the polymer jet is scattered as very fine fibers with a diameter at the nano-level [1–4]. The set-up of the electrospinning method is presented in **Figure 1**.

The polymers which are soluble in a solvent and have an enough high molecular weight to be electrospun can be used in nanofiber production by electrospinning. Many different natural polymers, synthetic polymers, and the polymer blends have been successfully electrospun into nanofibers by electrospinning method [12–21].

The main features that distinguish this method from other conventional methods are that the diameters of the fibers produced are below the micrometer range and it is possible to work with polymer solution or polymer melt [7, 22]. The application areas of the fibers produced by electrospinning method are quite extensive, and the material properties vary according to the application area.

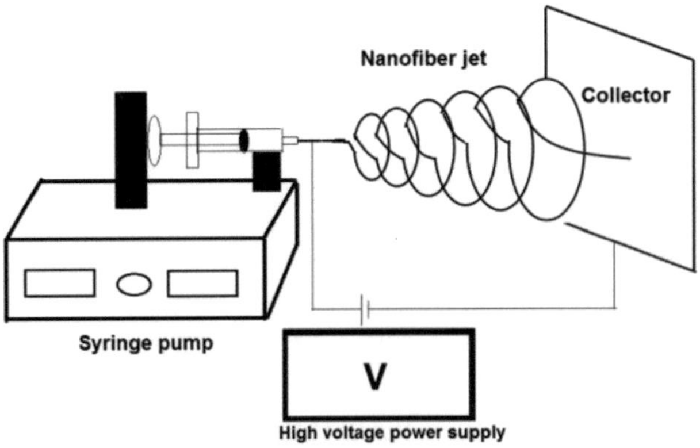

Figure 1. The set-up of electrospinning method.

http://dx.doi.org/10.5772/intechopen.74663

There are various parameters which affect the electrospinning process. These parameters can be divided into three groups as solution parameters, process parameters, and ambient parameters (**Table 1**) [1–3].

Nanofibers obtained by electrospinning method can be functionalized by loading some compounds such as drugs, dyes, metals, etc. to the polymer matrix [1, 5, 7–11]. Photochromic nanofiber, one of these types of functional nanofibers, is obtained by incorporating photochromic compounds into the polymer solution. Thus the nanofiber surface exhibits photochromic property, thereby increasing the different application possibilities.

Solution parameters	Process parameters	Ambient parameters
Concentration	Distance from nozzle to collector	Temperature
Molecular weight of the polymer	Flow rate	Humidity
Viscosity	Voltage	Atmospheric pressure
Electrical conductivity	Collector type	
Surface tension		

Table 1. The parameters affecting electrospinning process [1–3].

3. Photochromism

Photochromism is described as "the reversible absorption spectrum changes of a compound when the sample is irradiated with ultraviolet (UV) light" [23, 24]. Photochromic materials exhibit a reversible color change upon the change of UV light intensity. These compounds are generally colorless in the absence of UV light, while the compounds are colored with the increase in UV light intensity in the environment, and then the compound can return to its colorless state due to the decrease of UV light intensity. This photochromic behavior is stated as "positive photochromism." However, negative photochromism occurs if the compounds change their colored form into the colorless structure with UV light irradiation [25]. Photochromic materials may also be categorized into two groups as P-type and T-type according to the back reactions which occur with light (other wavelengths than UV) or heat in P-type and T-type photochromism, respectively [23, 26].

Photochromic compounds are widely used not only in optics but also in areas such as plastics, cosmetics, inks, and textiles [27]. The use of these compounds in textile field is problematic due to their poor affinity to the material, low solubility in water, sensitivity to high temperature, etc. [24, 28, 29]. Many different methods such as encapsulation, sol-gel processing, etc. have been developed to improve the application of photochromic compounds [24, 30–32]. Electrospinning can be seen as an alternative method for the use of photochromic compounds in textile field.

Application of photochromic compounds by electrospinning process provides advantages such as increasing sensitivity of photochromic compounds or reducing the time necessary for photochromic compound to respond to the UV irradiation due to large surface area of the electrospun mats [33]. Photochromic electrospun nanofibers could find applications in areas

Photochromic compounds	Photochromic reactions	Explanation of the substituent
Spiropyran	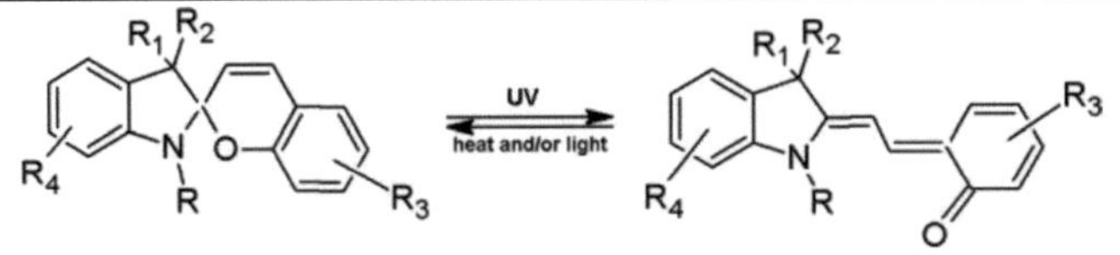	R: alkyl; R_1, R_2: alkyl (usually both methyl); R_3, R_4: H, halogen, nitro, etc.
Spirooxazine	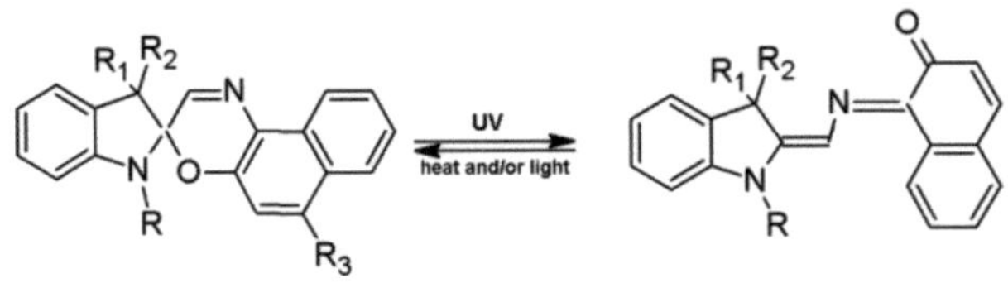	R: alkyl; R_1, R_2:alkyl (usually both methyl); R_3: H, halogen, etc.
Naphthopyran	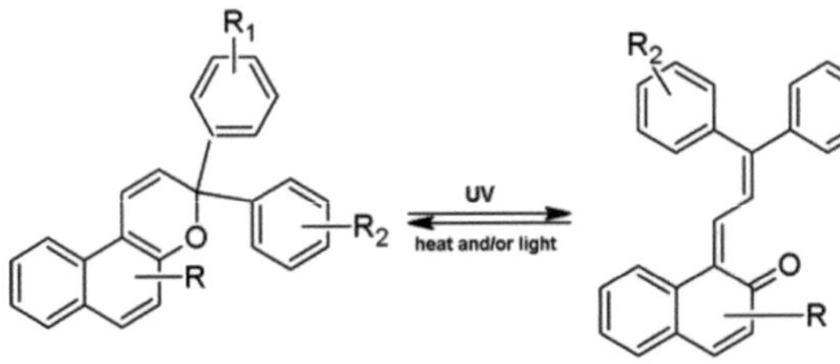	R: H, amino, etc.; R_1, R_2: H, alkoxy, amino, etc.
Diarylethene	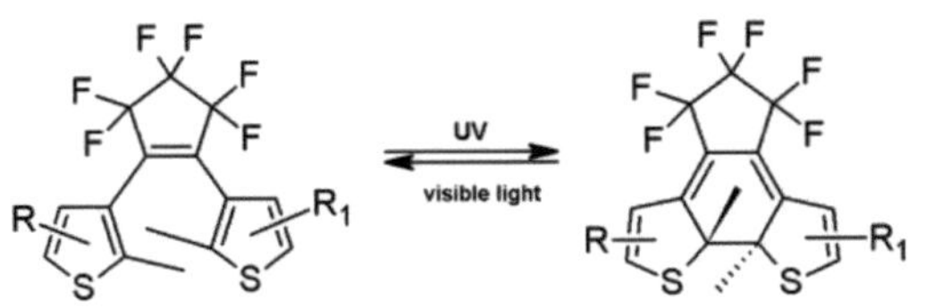	R, R_1: H, R_{1-2}, Ph, fused (het)aryl, etc.
Fulgide	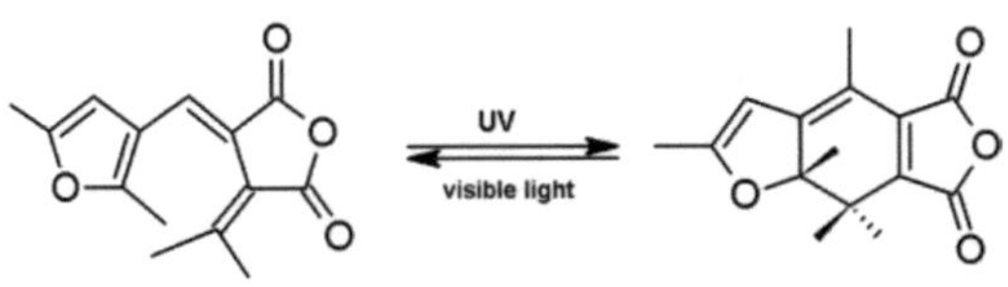	–
Azobenzene	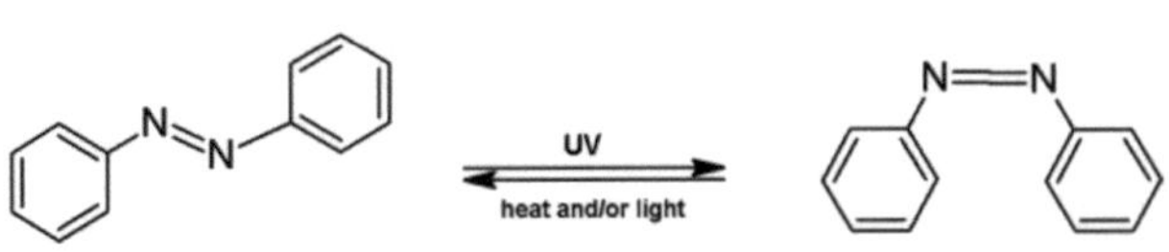	–
Stilbene	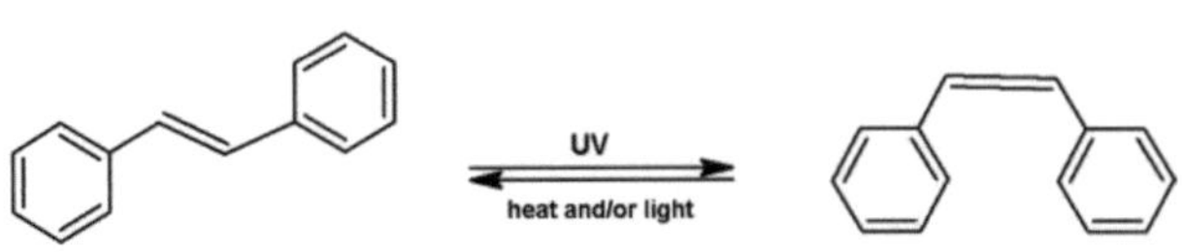	–

Table 2. Photochromic compounds and their photochromic reactions [23, 25, 26].

such as optical sensors, processing media, optical data storage devices, and functional components for smart surfaces [34, 35].

There are many different types of photochromic compounds as inorganic and organic. In the production of electrospun photochromic nanofiber, the organic photochromic compounds, such as spiropyran, spirooxazine, naphthopyrans, diarylethenes, and fulgides, which change their color based on pericyclic electrocyclic reactions, and azobenzene and stilbene, which change its color based on cis-trans isomerization, have been widely used (**Table 2**). Apart from such organic photochromic compounds, various different photochromic systems such as metal oxides have been also used in the production of electrospun photochromic nanofibers.

4. Photochromic nanofiber production

The production of electrospun photochromic nanofiber has been studied by a large number of authors. These studies which have used many different types of polymer and photochromic compound are discussed in depth below.

In the study of Akçakoca Kumbasar et al. [36], spirooxazine-based photochromic thermoplastic polyurethane (TPU) nanofiber mat, which is shown in **Figure 2**, has been obtained by electrospinning method. The photochromic coloration of the sample has occurred after UV irradiation (**Figure 2**). As shown in **Figure 3**, the neat nanofibers have been bead-free, and the inclusion of photochromic compounds in TPU polymer solution has not affected the smooth morphology of the resulting electrospun nanofibers (**Figure 3**).

Genovese et al. [34] have investigated the optochemosensing properties of the spiropyran (SP)-doped-poly(vinylidene fluoride-co-hexafluoropropylene) (PVDF-HFP) electrospun fibers. They have irradiated the SP-doped PVDF-HFP films, nanofibers, and fibers with UV light and then exposed to acidic vapors to determine both the photochromic and acidochromic properties of the samples. They have stated that the samples changed their color from colorless spiropyran form to red-colored merocyanine form with UV irradiation, and then, after the exposure to acidic vapors, the samples changed their red-colored structure to yellow protonated merocyanine form. In addition, the nanofibers have exhibited faster photochromism and protonation with respect to the fibers and films due to higher surface area of the nanofibers.

Durasevic [37] has reported the preparation of electrospun photochromic polyurethane (PU) fibers by incorporating the spirooxazine-based photochromic compound into the polymer solution. The samples have changed their colorless structure to deep blue color form with UV irradiation. However, this coloration has varied according to the UV light source type. The samples have exhibited darker color upon UVA irradiation with respect to the UVB irradiation. Durasevic [37] has specified that these electrospun photochromic materials can be used as UV sensor for medical textile application.

Khatri et al. [38] have prepared photochromic polyvinyl alcohol (PVA) nanofibers for the application of recording and erasing quick response (QR) codes. They have used spiropyran

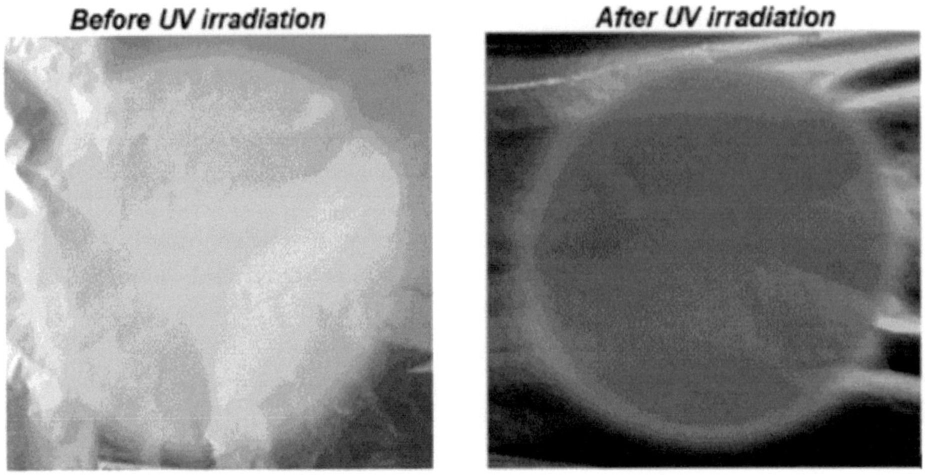

Figure 2. Images of the photochromic TPU nanofiber mat before and after UV irradiation [36].

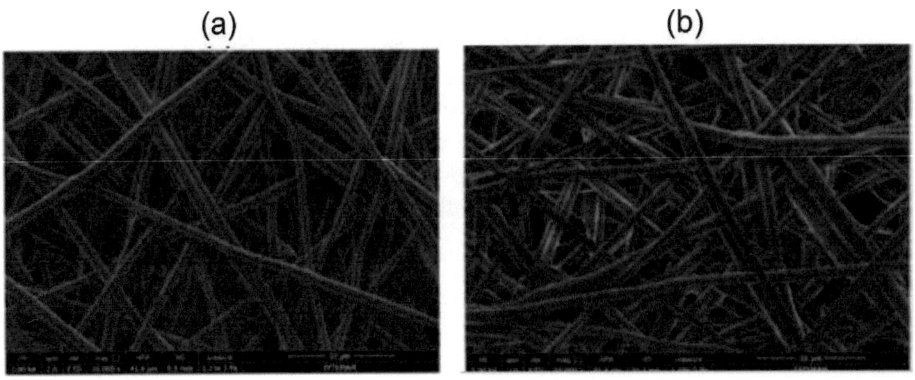

Figure 3. SEM images of (a) neat TPU nanofibers and (b) 20% photochromic compound-loaded TPU nanofibers [36].

and spirooxazine as photochromic compounds. They have stated that PVA is a useful polymer for the production of photochromic nanofibers due to the presence of OH groups in the polymer which leads to the formation of hydrogen bonding between the polymer and the photochromic compound, causing the uniform dispersion of the compounds within the nanofibers. They have also reported that the PVA-spiropyran nanofibers showed higher photocoloration and photo-reversibility than the PVA-spirooxazine nanofibers.

Liao et al. [35] have synthesized photoswitchable nanoparticles by grafting the fluorescent carbon nanoparticles with copolymers of styrene and spiropyran. Then, they have fabricated the nanofibers by electrospinning of these nanoparticles. The electrospun samples have exhibited reversibly photoswitchable fluorescence between blue-green and red colors.

In the study of Ali et al. [39], electrospun spiropyran-poly(ε-caprolactone) (PCL) fibers have been studied for patterned color image recordings. It has stated that coloration of the photochromic samples with UV irradiation has been faster than their decoloration. Their results have shown that the average fiber diameter of the neat fibers was larger with respect to spiropyran-PCL fibers. They have also performed the image recording on the electrospun photochromic fiber mat by write-erase-write technique with UV and visible lights.

Zillohu et al. [40] have investigated the structure and erasable writing properties of the electrospun spirooxazine-polyvinylidene fluoride (PVDF) fibers. Addition of spirooxazine compounds into the polymer matrix has made the fibers less polar and more amorphous. The authors have performed the pattern creating on the electrospun photochromic fibers by using water drops as lenses. They have also compared the UV absorption properties of the water drops loaded with silver and gold nanoparticles. The water drop containing silver nanoparticles has exhibited more absorption than the water drop loaded with gold particles.

The stability of electrospun naphthopyran-polyvinylpyrrolidone (PVP) nanofibers has been investigated in the study of Liu et al. [41]. They have used three different types of naphthopyran. The average diameters of naphthopyran-containing PVP fibers have been smaller with respect to the neat PVP fibers. The electrospun naphthopyran-polyvinylpyrrolidone (PVP) nanofibers have changed their colorless structure into the colored structure with UV irradiation after 5 min. However decoloration of the photochromic samples has been very slow. After removal of the samples from UV light, the color of the samples has remained even after 3 days. The authors have indicated that hydrogen bonding between PVP and naphthopyran caused the inhibition of naphthopyran fading.

Lee and Kim [42] have prepared the diarylethene-loaded polystyrene (PS) and polyacrylic acid (PAA) fibers by electrospinning method, and then they have studied on the image recording properties of the electrospun photochromic fibers. The diameter of the diarylethene-doped PAA fibers has been smaller than the diameter of the diarylethene-doped PS fibers, thereby increasing the resolution of patterned color images which were created on the electrospun mat upon photomasked UV irradiation. It has been found that these patterned color images have remained for 30 days.

Zhang et al. [43] have synthesized spiropyran-based PS nanowires and then used electrospinning method to obtain photochromic fibers. They have stated that the electrospun fibers have changed their color from white to pink under UV light.

In the study of Li et al. [33], spiropyran-based photochromic nanofiber mats have been obtained by electrospinning method. They have used poly(methyl methacrylate) (PMMA) and gelatin as the polymers. The hydrophilicity and photochromic properties of the samples have been investigated by the author. The water contact angle of spiropyran-gelatin nanofibers could have not been measured due to the water drop on the samples penetrated into them in 10 s. The water contact angles of spiropyran-PMMA nanofibers have decreased with UV irradiation as a result of the structural change of the photochromic compound from spiro to merocyanine form.

A photochromic spironaphthoxazine/isophorone-based fluorescent dye system has been developed by Lee et al. [44], and then they have used this dye system in the production of photochromic PMMA nanofiber by electrospinning method. Lee et al. [45] have also prepared the electrospun PMMA nanofiber-loaded photochromic spironaphthoxazine and D-π-A-type fluorescent dye. They have studied on the spectral and erasable writing properties of the photochromic samples at both studies. The authors have indicated that the electrospun photochromic PMMA nanofibers have a potential for optical data storage applications.

The hydrophilicity and photo-coloration properties of the electrospun spiropyran-PMMA and spiropyran-cellulose acetate (CA) nanofibers have been examined by Nammoonnoy et al. [46] and Shuiping et al. [47], respectively. They have reported that the photochromic nanofibers have turned from hydrophobic structure into hydrophilic structure by UV irradiation.

Lee et al. [48] have investigated the photocoloration of the electrospun spiropyran-loaded poly(ethylene oxide) (PEO) and polystyrene (PS) nanofibers. The electrospun spiropyran-loaded nanofibers have colored after photomasked UV irradiation, and then the samples have reverted back their original form with heat treatment at 120°C. However, after UV irradiation, the color of spiropyran-doped PEO nanofiber has been purple, while the color of spiropyran-doped PS nanofiber has been blue. The authors have indicated that spiropyrans tend to exhibit purple color in a polar environment and blue color in a nonpolar environment after UV irradiation. Thus, the spiropyran-loaded PEO nanofiber which is more polar than PS nanofiber has exhibited purple color after UV irradiation.

De Sousa et al. [49] have prepared spiropyran-cyclodextrin (SP-βCD) inclusion complexes and then they have electrospun SP-βCD and poly(methacrylic acid) (PMAA) blends into nanofibers. They have also fabricated spiropyran (SP)-PMAA nanofibers to analyze the effect of CD on photocoloration of spiropyran. The coloration rate of SP-βCD-PMAA nanofibers has been faster than the coloration rate of SP-PMAA nanofibers due to the hydroxyl groups of the βCD which lead to a stabilization of the merocyanine structure, in a more efficient way than those hydroxyl groups of PMAA polymer. In this study, water contact angles of the electrospun photochromic nanofibers have also been investigated. Incorporating SP-βCD inclusion complexes in the PMAA polymer matrix has increased the hydrophobicity of PMAA nanofibers. Additionally, the water contact angle of SP-βCD-PMAA nanofibers has decreased about 15° after UV irradiation.

In the study of Wang et al. [50], spiropyran-loaded poly(vinylidene fluoride-co-hexafluoropropylene) (PVDF-co-HFP) nanofiber mats have been fabricated by electrospinning method. They have obtained more uniform fibers upon addition of spiropyran to PVDF-co-HFP nanofiber due to increasing of the solution conductivity. The authors have stated that the spiropyran molecules have located near the core of the fibers according to the test results of XPS and water contact angle measurements.

Benedetto et al. [51] have reported the preparation and the characterization of electrospun photochromic PMMA fibers by incorporating the spiropyran-based photochromic compound into the polymer solution. They have specified that they used PMMA as polymer due to its high glass transition temperature, which supplies a more stable environment for the merocyanine form, thereby decelerating of the thermal decoloration rate. They have investigated the photocoloration and wettability properties of the electrospun photochromic nanofibers. It has

http://dx.doi.org/10.5772/intechopen.74663

been concluded that the electrospun spiropyran-loaded PMMA nanofibers has an opportunity to use as the light-driven nanometer-scale elements to be incorporated within optical interconnects, lab-on-a-chip technologies, and sensors.

Bianco et al. [52] have obtained diarylethene-loaded polyamide-6 nanofiber by electrospinning method and then analyzed morphology and photocoloration of the electrospun fibers. They have found a strong dichroism in the IR spectra of the diarylethene, which has confirmed the alignment of the diarylethene molecules with the main molecular axis along the fiber axis.

Gao et al. [53] have prepared photochromic fluorescence PVA nanofiber by electrospinning method. They have used three cyanostilbene derivatives as photochromic fluorescence compounds. The samples have changed their luminescence in different extents with UV irradiation for less than 1 min. The color of the fibers has been green before UV irradiation, while the color of the fibers has changed to cyan with UV irradiation. The authors have stated that the photochromic fluorescence PVA nanofibers have exhibited good reversibility and reproducibility, thereby showing potential for future practical sensor applications.

Bućko et al. [54] have synthesized azobenzene-based hybrid materials by sol-gel method and then used electrospinning method to obtain photochromic fibers. They have obtained the more beadles fibers with increasing the concentration of azo dyes in the sols. They have also measured the wettability of the samples to analyze the effect of the trans-cis isomerization on the water contact angle of the samples. The contact angle values of the samples have decreased with UV irradiation and have generally increased with increasing of electrospinning time and the dye content in the sols.

The electrospinning of azobenzene-cyclodextrin inclusion complex without using any polymer has been investigated in the study of Chen et al. [55]. They have also examined the UV response of the inclusion complexes before, during, and after electrospinning process. Before electrospinning process, the precipitation of azobenzene from the aqueous inclusion complex solution has occurred with UV irradiation. During the electrospinning process, UV irradiation has caused wider diameter distribution due to the interruption of inclusion complexes. After the electrospinning process, UV irradiation has modified the topography and adhesion forces of the electrospun nanofiber surfaces.

Photochromic superabsorber particles containing cross-linked hydrophilic core and hydrophobic azobenzene have been developed by Chen et al. [56], and then they have used various contents of these particles in the production of photochromic nanofiber by electrospinning method. Two different polymers as thermoplastic polyurethane (TPU) and polyamide (PA) have been used as carrier polymer matrix. The absorbency rate of the nanofibers has been fast; however, most of the photochromic superabsorber particles were released from the PA nanofibers after immersion in water for 24 h. The particle loss from the TPU nanofiber after third immersion cycles in water has been around 12 wt.%. It has been concluded that TPU was more stable matrix polymer for the particles with respect to PA. The absorbency capacity of the samples has increased with increasing of the photochromic superabsorber particle content in the nanofiber. The desorption rate of the TPU nanofibers has increased with UV irradiation due to the isomerization of the azobenzene compounds.

Organic photochromic compounds have been used in all of the studies described above. However, there are also photochromic nanofiber production studies using various different photochromic systems such as tungsten oxides.

Wei et al. [57] have prepared flexible rewritable nanofiber through electrospinning of tungsten oxide (WO_3)-loaded polyvinylpyrrolidone (PVP). The images have been written onto the electrospun samples with photomasked UV irradiation, and the time necessary for photochromic nanofiber to respond to the UV irradiation has been about 10 s. The erasure time of the image has been 1–2 days under ambient conditions, while the erasure time has been 20 min by heating to 80°C and 5 min by ozone treatment. The decoloration rate of the samples has also decelerated with increasing the WO_3 concentration in the fiber. The time necessary for the total decoloration has been extended to 10 days by adding 10% polyacrylonitrile (PAN) into the PVP matrix.

Zhang et al. [58] have developed PVP nanofiber mats consisted of tungsten-doped titanium dioxide (TiO_2) with different doping concentrations to mineralize toluene under visible light. The increasing of tungsten content in the nanofibers has increased the surface area while decreased the pore diameter. The mineralization degree of toluene has increased with increasing of tungsten content although increasing more than 20% tungsten concentration has caused decreased the mineralization degree due to the distortion of the anatase TiO_2 network.

Nguyen et al. [59] have synthesized mesoporous tungsten oxide nanofibers by electrospinning and surfactant-templated sol-gel process. Polyvinylpyrrolidone (PVP) has been used as polymer matrix. They have electrospun the WO_3-PVP nanofibers and then exposed the nanofibers to the heat treatment at 120°C overnight and then calcined at 500°C in air for 3 h. The nanofiber surfaces have become rough, and the diameters of the fibers have decreased after the heat treatments due to the calcination of organic polymer ingredients.

Jin et al. [60] have investigated the detection of volatile organic compound such as acetone by W-doped TiO_2-PVP nanofibers. The W-doped nanofibers have changed their colorless form to blue-colored form upon UV irradiation. The adsorption of acetone by the photochromic nanofibers has increased with UV irradiation, and this increase has become more apparent when the percentage of W was higher than 6%.

Unlike all these studies, Fischer and Hampp [61] have developed the photochromic PVA nanofibers with bacteriorhodopsin which is a biological photochromic pigment isolated from *Halobacterium salinarum*. They have studied on the process parameters for the electrospinning of bacteriorhodopsin-loaded PVA nanofibers.

All these studies of electrospun photochromic nanofiber production are listed in **Table 3**.

Photochromic compounds	Polymer matrices	Fiber diameter (nm)	References
Spiropyran	Poly(methyl methacrylate) (PMMA)	820–1430	Li et al. [33]
	Gelatin	340–870	
Spiropyran	Poly(vinylidene fluoride-cohexafluoropropylene) (PVDF-HFP)	800 ± 70	Genovese et al. [34]
Spiropyran	Polystyrene (PS)	500	Liao et al. [35]
Spirooxazine	Thermoplastic polyurethane (TPU)	700–900	Akcakoca Kumbasar et al. [36]
Spirooxazine	Polyurethane (PU)	610–4720	Durasevic [37]
Spiropyran	Polyvinyl alcohol (PVA)	307 ± 53	Khatri et al. [38]
Spirooxazine		349 ± 54	
Spiropyran	Poly(ε-caprolactone)	1130 ± 235	Ali et al. [39]
Spirooxazine	Polyvinylidene fluoride (PVDF)	1000 ± 200	Zillohu et al. [40]
Naphthopyran	Polyvinylpyrrolidone (PVP)	200–400	Liu et al. [41]
Diarylethene	Polystyrene (PS)		Lee and Kim [42]
	Polyacrylic acid (PAA)	800	
Spiropyran	Polystyrene (PS)	70	Zhang et al. [43]
Spironaphthoxazine/isophorone-fluorescent dye	Poly(methyl methacrylate) (PMMA)	400–1000	Lee et al. [44]
Spironaphthoxazine/D-p-A-type fluorescent dye (TCF)	Poly(methyl methacrylate) (PMMA)	400–1000	Lee et al. [45]
Spiropyran	Poly(methyl methacrylate) (PMMA)	—	Nammoonnoy et al. [46]
Spiropyran	Cellulose acetate (CA)	30–490	Shuiping et al. [47]
Spiropyran	Poly(ethylene oxide) (PEO)	—	Lee et al. [48]
	Polystyrene (PS)		
Spiropyran	Cyclodextrin molecule (βCD)	422 ± 40	De Sousa et al. [49]
	Poly(methacrylic acid) (PMAA)	526 ± 24	
Spiropyran	Poly(vinylidene fluoride-co-hexafluoropropylene) (PVDF-co-HFP)	2250 ± 190	Wang et al. [50]
Spiropyran	Poly(methacrylic acid) (PMAA)	50–400	Benedetto et al. [51]
Diarylethene	Polyamide 6	150	Bianco et al. [52]
Stilbene	Poly(vinyl alcohol) (PVA)	—	Gao et al. [53]

Photochromic compounds	Polymer matrices	Fiber diameter (nm)	References
Azobenzene	Triethoxyphenylsilane	300–1000	Bućko et al. [54]
Azobenzene	Hydroxypropyl-b-cyclodextrin (HPbCD)	460 ± 170–2320 ± 1140	Chen et al. [55]
Azobenzene/photochromic superabsorber particles	Thermoplastic polyurethane (TPU)	500–1000	Chen et al. [56]
	Polyamide (PA)		
Tungsten oxide (WO_3)	Polyvinylpyrrolidone (PVP)	200–1000	Wei et al. [57]
Tungsten-doped titanium dioxide (TiO_2)	Polyvinylpyrrolidone (PVP)	200–500	Zhang et al. [58]
Tungsten oxide (WO_3)	Polyvinylpyrrolidone (PVP)	233	Nguyen et al. [59]
Tungsten-doped titanium dioxide (TiO_2)	Polyvinylpyrrolidone (PVP)	800–900	Jin et al. [60]
Bacteriorhodopsin	Poly(vinyl alcohol) (PVA)	300–800	Fischer and Hampp [61]

Table 3. The studies of electrospun photochromic nanofiber production.

5. Conclusions

In this review, electrospun nanofiber production, functionalization of the nanofiber with photochromic compounds, and recent research development in this area have been described. Although the studies on photochromic nanofibers have been carried out, more detailed studies should be done on the production of the electrospun photochromic nanofibers, and alternative usage areas can be generated for these fibers.

Author details

Emriye Perrin Akçakoca Kumbasar[1]*, Seniha Morsunbul[1] and Simge Alır[2]

*Address all correspondence to: perrin.akcakoca@ege.edu.tr

1 Faculty of Engineering, Department of Textile Engineering, Ege University, Izmir, Turkey

2 Mayteks Örme Sanayi ve Ticaret A.Ş., Manisa, Turkey

References

[1] Huang ZM, Zhang YZ, Kotaki M, Ramakrishna S. A review on polymer nanofibers by electrospinning and their applications in nanocomposites. Composites Science and Technology. 2003;**63**(15):2223-2253. DOI: 10.1016/S0266-3538(03)00178-7

[2] Frenot A, Chronakis IS. Polymer nanofibers assembled by electrospinning. Current Opinion in Colloid and Interface Science. 2003;**8**(1):64-75. DOI: 10.1016/S1359-0294Ž03.0 0004-9

[3] Bhardwaj N, Kundu SC. Electrospinning: A fascinating fiber fabrication technique. Biotechnology Advances. 2010;**28**(3):325-347. DOI: 10.1016/j.biotechadv.2010.01.004

[4] Ramakrishna S. An Introduction to Electrospinning and Nanofibers. Singapore: World Scientific Publishing Company; 2005. 382 p. DOI: 10.1142/5894

[5] Luo Y, Nartker S, Miller H, Hochhalter D, Wiederoder M, Wiederoder S, et al. Surface functionalization of electrospun nanofibers for detecting E. coli O157:H7 and BVDV cells in a direct-charge transfer biosensor. Biosensors and Bioelectronics. 2010;**26**(4):1612-1617. DOI: 10.1016/j.bios.2010.08.028

[6] Kim C, Yang KS. Electrochemical properties of carbon nanofiber web as an electrode for supercapacitor prepared by electrospinning. Applied Physics Letters. 2003;**83**(6):1216-1218. DOI: 10.1063/1.1599963

[7] Agarwal S, Wendorff JH, Greiner A. Use of electrospinning technique for biomedical applications. Polymer. 2008;**49**(26):5603-5621. DOI: 10.1016/j.polymer.2008.09.014

[8] Sill TJ, Von Recum HA. Electrospinning: Applications in drug delivery and tissue engineering. Biomaterials. 2008;**29**(13):1989-2006. DOI: 10.1016/j.biomaterials.2008.01.011

[9] Tidjarat S, Winotapun W, Opanasopit P, Ngawhirunpat T, Rojanarata T. Uniaxially aligned electrospun cellulose acetate nanofibers for thin layer chromatographic screening of hydroquinone and retinoic acid adulterated in cosmetics. Journal of Chromatography A. 2014;**1367**:141-147. DOI: 10.1016/j.chroma.2014.09.043

[10] Hu X, Liu S, Zhou G, Huang Y, Xie Z, Jing X. Electrospinning of polymeric nanofibers for drug delivery applications. Journal of Controlled Release. 2014;**185**:12-21. DOI: 10.1016/j.jconrel.2014.04.018

[11] Ner Y, Grote JG, Stuart JA, Sotzing GA. White luminescence from multiple-dye-doped electrospun DNA nanofibers by fluorescence resonance energy transfer. Angewandte Chemie International Edition. 2009;**48**(28):5134-5138. DOI: 10.1002/anie.200900885

[12] Stephens JS, Chase DB, Rabolt JF. Effect of the electrospinning process on polymer crystallization chain conformation in Nylon-6 and Nylon-12. Macromolecules. 2004;**37**(3):877-881. DOI: 10.1021/ma0351569

[13] Deitzel JM, Kleinmeyer JD, Hirvonen JK, Tan NCB. Controlled deposition of electrospun poly(ethylene oxide) fibers. Polymer. 2001;**42**(19):8163-8170. DOI: 10.1016/S0032-3861(01)0 0336-6

[14] Koski A, Yim K, Shivkumar S. Effect of molecular weight on fibrous PVA produced by electrospinning. Materials Letters. 2004;**58**(3):493-497. DOI: 10.1016/S0167-577X(03)00532-9

[15] Choi SS, Lee YS, Joo CW, Lee SG, Park JK, Han KS. Electrospun PVDF nanofiber web as polymer electrolyte or separator. Electrochimica Acta. 2004;**50**(2):339-343. DOI: 10.1016/j.electacta.2004.03.057

[16] You Y, Min BM, Lee SJ, Lee TS, Park WH. In vitro degradation behavior of electrospun polyglycolide, polylactide, and poly(lactide-coglycolide). Journal of Applied Polymer Science. 2005;**95**(2):193-200. DOI: 10.1002/app.21116

[17] Ayutsede J, Gandhi M, Sukigara S, Micklus M, Chen HE, Ko F. Regeneration of *Bombyx mori* silk by electrospinning. Part 3: Characterization of electrospun nonwoven mat. Polymer. 2005;**46**(5):1625-1634. DOI: 10.1016/j.polymer.2004.11.029

[18] Hadipour-Goudarzi E, Montazer M, Latifi M, Aghaji AAG. Electrospinning of chitosan/sericin/PVA nanofibers incorporated with in situ synthesis of nano silver. Carbohydrate Polymers. 2014;**113**:231-239. DOI: 10.1016/j.carbpol.2014.06.082

[19] Bak SY, Yoon GJ, Lee SW, Kim HW. Effect of humidity and benign solvent composition on electrospinning of collagen nanofibrous sheets. Materials Letters. 2016;**181**:136-139. DOI: 10.1016/j.matlet.2016.06.019

[20] Akduman C, Özgüney I, Kumbasar EPA. Preparation and characterization of naproxen-loaded electrospun thermoplastic polyurethane nanofibers as a drug delivery system. Materials Science and Engineering: C. 2016;**64**:383-390. DOI: 10.1016/j.msec.2016.04.005

[21] Dai J, Yang S, Jin J, Li G. Electrospinning of PLA/pearl powder nanofibrous scaffold for bone tissue engineering. RSC Advances. 2016;**6**(108):106798-106805. DOI: 10.1039/C6RA21796F

[22] Greiner A, Wendorff JH. Electrospinning: A fascinating method for the preparation of ultrathin fibers. Angewandte Chemie International Edition. 2007;**46**(30):5670-5703. DOI: 10.1002/anie.200604646

[23] Crano JC, Guglielmetti RC. Organic Photochromic and Thermochromic Compounds, Main Photochromic Families. Vol. 1. New York: Kluwer Academic Publishers; 2002. 378 p. DOI: 10.1007/b114211

[24] Akçakoca Kumbasar EP, Çay A, Morsunbul S, Voncina B. Color build-up and UV-protection performance of encapsulated photochromic dye-treated cotton fabrics. AATCC Journal of Research. 2016;**3**(2):1-7. DOI: 10.14504/ajr.3.2.1

[25] Xiaoming Tao, editor. Handbook of Smart Textiles. Singapore: Springer Science+Business Media; 2015. 1063 p. DOI: 10.1007/978-981-4451-45-1

[26] Corns SN, Partington SM, Towns AD. Industrial organic photochromic dyes. Coloration Technology. 2009;**125**(5):249-261. DOI: 10.1111/j.1478-4408.2009.00204.x

[27] Ferrara M, Bengisu M. Materials That Change Color: Smart Materials, Intelligent Design. Heidelberg: SpringerBriefs in Applied Sciences and Technology; 2014. 139 p. DOI: 10.1007/978-3-319-00290-3

[28] Billah SMR, Christie RM, Shamey R. Direct coloration of textiles with photochromic dyes. Part 1: Application of spiroindolinonaphthoxazines as disperse dyes to polyester, nylon and acrylic fabrics. Coloration Technology. 2008;**124**(4):223-228. DOI: 10.1111/j.1478-4408.2008.00145.x

[29] Little AF, Christie RM. Textile applications of photochromic dyes. Part 2: Factors affecting the photocoloration of textiles screen-printed with commercial photochromic dyes. Coloration Technology. 2010;**126**(3):164-170. DOI: 10.1111/j.1478-4408.2010.00242.x

[30] Feczkóa T, Samu K, Wenzel K, Neral B, Voncina B. Textiles screen-printed with photochromic ethyl cellulose–spirooxazine composite nanoparticles. Coloration Technology. 2012;**129**(1):18-23. DOI: 10.1111/j.1478-4408.2012.00404.x

[31] Mahltig B, Textor T, Akcakoca KP. Photobactericidal and photochromic textile materials realized by embedding of advantageous dye using sol-gel technology. Celal Bayar University Journal of Science. 2015;**11**(3):306-315. DOI: 10.18466/cbujos.44647

[32] Cheng T, Lin T, Fang J, Brady R. Photochromic wool fabrics from a hybrid silica coating. Textile Research Journal. 2007;**77**(12):923-928. DOI: 10.1177/0040517507083523

[33] Li X, Lin L, Kanjwal MA, Chronakis IS, Liu S, Chen Y. Preparing photochromic nanofibers and animal cells using a photochromic compound of 1,3,3-trimethyl-6-nitrospiro (2H-1-benzopyran-2,2-indoline). Colloids Surfaces B Biointerfaces. 2012;**89**:67-72. DOI: 10.1016/j.colsurfb.2011.08.024

[34] Genovese ME, Colusso E, Colombo M, Martucci A, Athanassiou A, Fragouli D. Acidochromic fibrous polymer composites for rapid gas detection. Journal of Materials Chemistry A. 2017;**5**:339-348. DOI: 10.1039/C6TA08793K

[35] Liao B, Wang W, Long P, Deng X, He B, Liu Q, Yi S. The carbon nanoparticles grafted with copolymers of styrene and spiropyran with reversibly photoswitchable fluorescence. Carbon. 2015;**91**:30-37. DOI: 10.1016/j.carbon.2015.04.030

[36] Akçakoca Kumbasar EP, Morsunbul S, Akduman Ç, Alır S. Preparation of photochromic thermoplastic polyurethane fibers by electrospinning. In: 8th International Textile, Clothing & Design Conference—Magic World of Textiles; October 2-5; Dubrovnik, Croatia. 2016.

[37] Durasevic V. Smart dyes for medical textiles. In: Van Langenhove L, editor. Advances in Smart Medical Textiles: Treatments and Health Monitoring. Woodhead Publishing Series in Textiles. 1st ed. Cambridge: Woodhead Publishing; 2016. pp. 19-55. DOI: 10.1016/B978-1-78242-379-9.00002-5

[38] Khatri Z, Ali S, Khatri I, Mayakrishnan G, Hun S, Kim I. UV-responsive polyvinyl alcohol nanofibers prepared by electrospinning. Applied Surface Science. 2015;**342**:64-68. DOI: 10.1016/j.apsusc.2015.03.046

[39] Ali S, Ahmed F, Khatri Z, Kim SH. Fabrication of rewritable pattern images with ultraviolet-responsive polycaprolactone electrospun fibers. Textile Science and Engineering. 2015;**52**:126-131. DOI: 10.12772/TSE.2015.52.126

[40] Zillohu AU, Abdelaziz R, Homaeigohar S, Krasnov I, Müller M, Strunskus T, Elbahri M. Biomimetic transferable surface for a real time control over wettability and photoerasable writing with water drop lens. Scientific Reports. 2014;**4**(7407):1-7. DOI: 10.1038/srep07407

[41] Liu L, Zhang Y, Wang G, Zhang M, Guan H. High stability of photoinduced merocyanine in naphthopyran-doped polyvinylpyrrolidone electrospun nanofibers. Polymer International. 2014;**63**:1991-1996. DOI: 10.1002/pi.4750

[42] Lee J, Kim J. Rewritable color patterns with photochromic diarylethene-embedded electrospun fibers. Bulletin of the Korean Chemical Society. 2014;**35**:3381-3384. DOI: 10.5012/bkcs.2014.35.11.3381

[43] Zhang W, Hong C, Pan C. Fabrication of electrospinning fibers from spiropyran-based polymeric nanowires and their photochromic properties. Macromolecular Chemistry Physics. 2013;**214**:2445-2453. DOI: 10.1002/macp.201300428

[44] Lee E, Gwon S, Ji B, Wang S, Kim S. Photoswitching electrospun nano fiber based on a spironaphthoxazine e isophorone-based fluorescent dye system. Dyes and Pigments. 2011;**92**:542-547. DOI: 10.1016/j.dyepig.2011.06.002

[45] Lee E, Gwon S, Son Y, Kim S. Modulation of a fluorescence switch of nanofiber mats containing photochromic spironaphthoxazine and D-π-A charge transfer dye. Journal of Luminescence. 2012;**132**:1427-1431. DOI: 10.1016/j.jlumin.2012.01.034

[46] Nammoonnoy J, Koesdjojo M T, Frederick R T, Remcho V T. Photo-actuated droplet microfluidics. In: International Conference on Miniaturized Systems for Chemistry and Life Sciences (μTAS '15); October 2-6, 2011; Seattle, Washington, USA. 2011. p. 933-935

[47] Shuiping L, Lianjiang T, Weili H, Xiaoqiang L, Yanmo C. Cellulose acetate nanofibers with photochromic property: Fabrication and characterization. Materials Letters. 2010;**64**:2427-2430. DOI: 10.1016/j.matlet.2010.08.018

[48] Lee J, Lee C, Kim J. Fabrication of patterned images in photochromic organic microfibers. Macromolecular Rapid Communications. 2010;**31**:1010-1014. DOI: 10.1002/marc.201000019

[49] De Sousa BF, Guerreiro DTJ, Ma M, Anderson DG, Drum CL, Sinisterra DR, Langer R. Photo-response behavior of electrospun nanofibers based on spiropyran-cyclodextrin modified polymer. Journal of Materials Chemistry. 2010;**20**:9910-9917. DOI: 10.1039/c0jm01903h

[50] Wang M, Vail SA, Keirstead AE, Marquez M, Gust D, Garcia AA. Preparation of photochromic poly (vinylidene fluoride-co-hexafluoropropylene) fibers by electrospinning. Polymer. 2009;**50**:3974-3980. DOI: 10.1016/j.polymer.2009.06.044

[51] Benedetto DF, Mele E, Camposeo A, Athanassiou A, Cingolani R, Pisignano D. Photoswitchable organic nanofibers. Advanced Materials. 2008;**20**:314-318. DOI: 10.1002/adma.200700980

[52] Bianco A, Iardino G, Manuelli A, Bertarelli C, Zerbi G. Strong orientation of polymer chains and small photochromic molecules in polyamide 6 electrospun fibers. Chemphyschem. 2007;**8**:510-514. DOI: 10.1002/cphc.200600657

[53] Gao R, Cao D, Guan Y, Yan D. Flexible self-supporting nanofibers thin films showing reversible photochromic fluorescence. American Chemical Society Applied Materials & Interfaces. 2015;**7**:9904-9910. DOI: 10.1021/acsami.5b01996

[54] Bućko A, Zielińska S, Ortyl E, Larkowska M, Barille R. Synthesis of organic–inorganic hybrid azobenzene materials for the preparation of nanofibers by electrospinning. Optical Materials. 2014;**38**:179-187. DOI: 10.1016/j.optmat.2014.10.021

[55] Chen M, Nielsen SR, Uyar T, Zhang S, Zafar A, Dong M, Besenbacher F. Electrospun UV-responsive supramolecular nanofibers from a cyclodextrin–azobenzene inclusion complex. Journal of Materials Chemistry C. 2013;**1**:850-855. DOI: 10.1039/c2tc00180b

[56] Chen F, Greiner A, Agarwal S. Stimuli-responsive elastic polyurethane-based superabsorber nanomat composites. Macromolecular Materials and Engineering. 2011;**296**:517-523. DOI: 10.1002/mame.201000387

[57] Wei J, Jiao X, Wang T, Chen D. Electrospun photochromic hybrid membranes for flexible rewritable media. American Chemical Society Applied Materials & Interfaces. 2016;**8**:29713-29720. DOI: 10.1021/acsami.6b10620

[58] Zhang L, Li Y, Xie H, Wang H, Zhang Q. Efficient mineralization of toluene by W-doped TiO_2 nanofibers under visible light irradiation. Journal of Nanoscience and Nanotechnology. 2015;**15**:2944-2951. DOI: 10.1166/jnn.2015.9638

[59] Nguyen TA, Jun TS, Rashid M, Kim YS. Synthesis of mesoporous tungsten oxide nanofibers using the electrospinning method. Materials Letters. 2011;**65**:2823-2825. DOI: 10.1016/j.matlet.2011.05.103

[60] Jin M, Zhang X, Pu H, Nishimoto S, Murakami T, Fujishima A. Photochromism-based detection of volatile organic compounds by W-doped TiO_2 nanofibers. Journal of Colloid and Interface Science. 2011;**362**:188-193. DOI: 10.1016/j.jcis.2011.06.041

[61] Fischer T, Hampp NA. Encapsulation of purple membrane patches into polymeric nanofibers by electrospinning. IEEE Transactions on Nanobioscience. 2004;**3**:118-120. DOI: 10.1109/TNB.2004.828220

Chapter 5

Electrospun Bead-on-String Fibers: Useless or Something of Value?

Huijing Zhao and Huanjie Chi

Additional information is available at the end of the chapter

http://dx.doi.org/10.5772/intechopen.74661

Abstract

Bead-on-string fibers, which were initially thought to be a "by-product" of the electrospun fibers, are widely observed in electrospinning, which is a convenient method to produce nanofibers. The electrospun bead-on-string fibers were thought to have detrimental properties and were generally discarded, but recently they have gained attention since they are considered to have promising applications in many fields, including tissue engineering, drug delivery, and air/water filtration, among others. This chapter is a comprehensive and systematic literature review that summarizes the processes, methods, vital influencing factors, formation conditions, morphology changes, and applications of the electrospun bead-on-string fibers. It helps to understand the current research status and to further understand the mechanism by which these bead-on-string fibers are formed.

Keywords: electrospinning, bead-on-string fibers, instability, filtration, sustained drug

1. Introduction

Electrospinning is a powerful and effective technique to fabricate nanoscale fibers from polymer solutions or melts [1]. The electrospun nanofibers have been widely used as scaffolds in tissue engineering, drug delivery, industrial filter material, wound dressing, and composite applications owing to their porous and high specific surface area [2–6]. In 1934, Furmhals patented the experimental device for the preparation of polymer fibers via electrostatic power which is considered as the beginning of electrospun nanofibers [7]. Following several decades since the invention of experimental device, the development of electrostatic spinning was

IntechOpen

© 2018 The Author(s). Licensee IntechOpen. This chapter is distributed under the terms of the Creative Commons Attribution License (http://creativecommons.org/licenses/by/3.0), which permits unrestricted use, distribution, and reproduction in any medium, provided the original work is properly cited.

attended to improve the setup for producing polymer filaments [8–10]. However, recently, electrospinning technology has received wide attention together with the development of nanotechnology [11]. Electrostatic spinning technology has gradually become the most direct and effective method to prepare nanofiber materials.

Bead-on-string fibers have been widely observed in electrospun products. At first, bead-on-string fibers were considered as useless products in electrospinning which could affect the performance of nanomaterials since the beads would greatly reduce the surface area [12]. However, further studies showed that the incorporated beads in the micron range are effective for drug-loading purposes [13]. In this sense, spheres are able to solve the problem of having electrospun fibers that are too fine to incorporate high doses of drug in the tissue engineering scaffolds [14, 15]. The particular structure of bead-on-string fibers, mainly composed of micron-sized spheres and nanometer-sized fibers, presents some microporosity that can be used in air/water filter to solve the problem of low filtration efficiency and high air pressure resistance of highly efficient filtration materials [16–18]. In summary, bead-on-string fibers have potential applications in tissue engineering, drug delivery, as well as in air and water filters.

In recent years, the development of bead-on-string fibers has received a lot of attention. However, most studies focused on the influential factors of the formation of electrospun bead-on-string fibers [19, 20]. Until now, there have been few reviews about electrospun bead-on-string fibers. This chapter provides a detailed and systematic description on the processes, methods, influential factors, formation conditions, morphology changes, and applications of the electrospun bead-on-string fibers, which is based on in-depth literature survey of this technique in recent decades.

2. Preparation methods of electrospun bead-on-string fibers and the polymers used

Traditional electrospinning devices were mainly composed of high-voltage power supply (a few thousand voltages to tens of thousands of voltages), solution storage device (polymer solution or melt), spraying device (capillary needle with millimeter level) and collecting device (metal plate or aluminum foil) [10]. In order to overcome the low production efficiency, new spinning equipment have recently appeared, such as multi-nozzle electrostatic spinning, non-nozzle electrostatic spinning, and bubble electrospinning, which laid the foundation for the industrialization of the electrospinning nanofibers [21–23].

The fundamental principle of electrospinning is that the solution forms a liquid drop at the end of the spraying device under a high-voltage electric field. As the voltage increases, the shape of the droplet gradually changes to form the Taylor cone [9, 12]. When the electric field force of the solution or the melt overcomes the surface tension and the viscous force, the droplet of the polymer solution or melt is ejected from the tip of the needle in the form of a jet and solidified on the collecting device [24].

2.1. Traditional electrospinning

Almost all electrospinning methods can be used to produce bead-on-string fibers by adjusting the processing parameters. Although the electrospinning methods and devices used to prepare bead-on-string fibers have some differences, the working principle of electrostatic spinning and formation mechanism of bead-on-string fibers are the same.

Traditional electrospinning is the most common method to prepare bead-on-string fibers [25]. However, the polymer used in traditional electrospinning must be soluble in the solvent. Therefore, it is difficult to use polymers that cannot be dissolved in the solvent to form bead-on-string fibers. With the wide application of electrospun bead-on-string fibers in drug delivery, emulsion electrospinning and coaxial electrospinning have also emerged as interesting methods for preparing bead-on-string fibers [26].

2.2. Emulsion electrospinning

Polymers that do not have tendency to form nanofibers from its solution or melt can be dispersed in a carrier that can easily form nanofibers in order to prepare nanofibers by emulsion electrospinning [27]. In this way, water-soluble drugs can be dispersed inorganic solutions of polymers to form stable homogeneous oil in water (O/W) or water in oil (W/O) emulsions. Large droplets of O/W emulsion will burst into small droplets in the electrospinning process. Bead-on-string fibers can be obtained [28, 29]. However, the W/O emulsion is different. The formation mechanism of bead-on-string fibers is based on droplets in aqueous phase that move to the fiber center and combine, which is followed by the jet stretching and solvent evaporation, which cause demulsification, allowing the formation of the bead-on-string fibers [30]. During the preparation of drug-loaded bead-on-string fibers by the emulsion electrospinning, the drugs and polymers are not necessary to be solved in the same solvent. Furthermore, the water soluble drugs can be formed as floating droplets in the emulsions and embedded in the beads once the bead-on-string fibers are prepared. Qi et al. successfully prepared bead-on-string fibers and performed sustained release of drugs by adding calcium chloride in alginate aqueous solution to crosslink Ca ions and alginate to form calcium alginate microspheres encapsulating drug model Bovine serum albumin (BSA) [26].

2.3. Coaxial electrospinning

Nozzles for coaxial electrospinning consist of two capillary tubes, including the inner layer and the outer layer capillary tubes. The principle of coaxial electrospinning is that two kinds of solutions are injected into the inner and outer capillary tubes to join at nozzle tip. Before the solution is sufficiently mixed, the mixture is sprayed out to form a stratified polymer jet since the large electric field force droplets are subjected at the end of the spraying device, allowing solvent evaporation and the solution solidification, forming a core-sheath structure obtained on the collecting device [31, 32]. If the drug or the bioactive factor is encapsulated in the polymer cortex, the nanofibers can be used as a drug delivery carrier [32–35].

Coaxial electrospinning can overcome the shortcomings of uneven distribution of drugs due to poor spinnability of drugs or poor compatibility of drugs and polymers. Furthermore, durgs to be loaded are not necessary to be spinnable [36]. Tian et al. successfully prepared bead-on-string fibers by coaxial electrospinning. In this experiment, the outer fluid was a 20% (w/v) poly(ethylene glycol) (PEG) (M = 20,000) solution in a mixed solvent of N,N-dimethylformamide (DMF) and methylene chloride (MC v/v = 1:1), and the inner fluid was a 35% (w/v) polystyrene (PS, M_W = 350,000) solution in DMF. The study also showed that the bead region and the fiber part were constituted by outer fluid and inner fluid, respectively, through energy-dispersive X-ray spectroscopy [37].

Using different methods of preparation as mentioned above, there are a bunch of polymers that can be used to produce bead-on-string fibers, as shown in **Table 1**. The solvents and some of the additives that can promote the formation of bead-on-string fibers are listed.

Polymer	Solvent	Additives	References
Poly(hydroxybutyrate-co-valerate) (PHBV)	Chloroform (CF)	—	[38, 39]
Poly(ethylene oxide) (PEO)	Distilled water	—	[25, 30]
PEG + PS	DMF + MC	—	[37]
Poly (L-lactic acid) (PLLA)	Dichloromethane (DCM)	Alginate, sodium bis (2-ethylhexyl) sulfosuccinate	[26]
Poly-(methyl methacrylate) (PMMA)	DMF	Titanium, tetrachloride ($TiCl_4$) hydrolyzed nanoparticles	[40]
Poly (butylene succinate) (PBS)	CF, DCM, 2-chloroethanal (CE), isopropanol (IPA)	LiCl	[41]
Silk fibroin + gelatin	Formic acid	—	[42]
PS	1,2-Dichloroethane, DMF, ethylacetate (EA), methylethylketone (MEK), Tetrahydrofuran (THF)	—	[19, 43]
Chitosan	Glacial acetic acid + deionized water	—	[44]
Poly (Ɛ-caprolactone) (PCL) + gelatin	Acetic acid + EA + water	—	[4]
Poly (lactic acid) (PLA)	N,N-dimethylacetamide (DMAC) + DCM	—	[45]
Silk-like polymer with fibronectin functionality (SLPF)	Formic acid	—	[46]
Poly (vinyl alcohol) (PVA)	Distilled water	Ethanol, NaCl	[47]
Polyvinylpyrrolidone (PVP)	Ethanol + water	LiCl, NaCl, $MgCl_2$	[48]

Table 1. Polymers used for producing bead-on-string fibers.

3. Influencing factors of electrospun bead-on-string fibers and their formation conditions

Electrospinning is a process in which the polymer solution forms a Taylor cone at the end of the spraying device under high electrostatic voltage. In parallel, there is competition between viscous resistance, surface tension, and electric field force suffered by the polymer solution. When the electric field force reaches a certain critical value, the liquid drops overcome the effect of the viscous resistance and surface tension, and are then ejected from the tip of the needle in the form of jet. Finally, the jet is solidified to form fibers on the collecting device [9, 24, 49]. In the initial stage of jet sprayed from the tip, the path of the jet has linear acceleration motion. Due to the existence of the viscous resistance, the acceleration decreases continuously, when the acceleration is zero or constant, making the presence of any disturbance change the linear motion of the jet forming the instability of the jet [24, 50, 51].

Instability of jet is a transport phenomenon where it may be originated from a particular disturbance or fluctuation and extended at different rates with time. There are three modes of instabilities in electrospinning process. The first one is the Rayleigh mode, which is the axisymmetric extension of the classical Rayleigh instability when electrical effects are important, and then the axisymmetric conducting mode, and the whipping conducting mode. The latter are dubbed "conducting modes" because they are sensitive to the surface and conductivity of electrospinning solution at high electric field [51, 52]. Among three types of jet instability that are related to the instability modes, the Rayleigh instability is driven by the surface tension, the varicose instability and whipping instability are caused by the nature of electricity. One or more different instability modes may occur in electrospinning process, which depends on the basic processing parameters such as the velocity, radius, and surface charge density of the jet [53–55].

The formation of beads is related to the three types of instability in electrospinning process. It is generally accepted that Rayleigh instability and varicose instability are favorable for the formation of beads, while whipping instability inhibits bead formation [38, 51, 52]. In electrospinning process, jet instability caused by capillary waves will lead to burst of cylindrical jet and then breakup of liquid droplet gradually shrinks into the ball to get the smallest surface area due to surface tension [24]. Owing to the existence of instability, spherical small droplets produce deformation, which will lead to an increase in the surface area and surface energy, which causes instabilities of jet to continue. Finally, bead-on-string fibers are collected on the collection device [25]. Similarly, varicose instabilities are caused by the nature of electricity in which the charges on the jet repel each other and jet is stretched and refined under the force of an electric field. When the charges on jet suddenly decrease, cylindrical jet will collapse to form droplets and will be stretched due to the charge repulsion. Ultimately, bead-on-string fibers are formed [56, 57]. Whipping instability is caused by the excessive charges on the surface of cylindrical jet which can generate fluctuation. When the fluctuation is big enough, jet will split into many small branches. If the whip continues to exist, the jet will split into smaller branches to form nanoscale fibers [24]. When the Rayleigh instability and varicose instability dominate jet instabilities, bead-on-string fibers will form. If the whipping instability is dominant in jet instabilities, smooth fibers will be obtained [56–59].

The aforementioned analysis intends to show which is in a dominant position among the Rayleigh instability, varicose instability and whipping instability determines whether bead-on-string fibers can be formed. Surface tension, viscosity, and charge density of spinning solution have a great impact on the three instabilities above. Therefore, influencing factors of bead-on-string fiber formation can be discussed from three aspects of surface tension, viscosity, and charge density of the spinning solution.

3.1. Surface tension of spinning solution

The surface tension of solution is the cohesive force between the molecules of the same kind, which makes the surface molecules of the liquid close to each other, and the liquid shows automatic shrinking to obtain the minimum surface area. Rayleigh instability, one of axisymmetric instability, is driven by surface tension that the vertical jet often suffers by surface tension to form a symmetrical waveform in the free-falling process [60, 61]. Therefore, surface tension has an important effect on the formation of bead-on-string fibers [62]. In recent years, there are many researchers working on the effect of surface tension on the formation of bead-on-string fibers. Most works have focused on the effect of the surface tension on the formation of bead-on-string fibers by changing the type of solvent or the mixing ratio of spinning solutions and comparing the morphology of fibers. Fong et al. studied the variation of water and alcohol ratio to obtain different surface tension of poly(ethylene oxide) (PEO) solutions, showing that the increase in surface tension of polymer solution was favorable to the formation of bead-on-string fibers [25]. Zuo et al. and his colleagues studied the effect of surface tension of added PHBV solution on the formation of bead-on-string fibers. The results showed that greater surface tension of spinning solution was more likely to obtain bead-on-string fibers. The reason is that the surface tension of the polymer solution attempts to obtain smaller mass unit area by shrinking jet into a sphere when the jets are broken, which leads to the increase of Rayleigh instability and the formation of bead-on-string fibers [38]. Therefore, it is reasonable that controlling surface tension of spinning solution is crucial for the preparation of the desired bead size and morphology.

3.2. Viscosity of spinning solution

The viscosity of the solution determines the ability of a solution to flow under external force, where the internal friction force is generated at the interface due to the flow of solution being limited by entanglement of molecular chains. This property of polymer solution is characterized by viscosity. All factors that cause the increase of entanglement density can make the movement of molecular chain more difficult and increase the viscosity of solution. The number of molecular weight per unit volume characterized by the concentration of the solution and the length of the molecular chain can affect the viscosity of solution and hence the formation of bead-on-string fibers [63]. Generally speaking, the greater the concentration and molecular weight of solution, the stronger the entanglement degree and the cohesion molecular chains, which leads to a greater concentration of spinning solution. Many research works have indicated that the greater the viscosity is, the stronger viscous resistance the jet has, which is more conducive to resist surface tension. In this case, the Rayleigh instability and varicose instability are limited, and the whipping instability is dominant in the jet instability, inhibiting the formation of the beads [53, 64–66]. There are many studies about the effects of molecular weight, viscosity, and concentration of polymer solution on the formation of bead-on-string fibers. Gupta et al. explored relationships between fiber formation, viscosity, molecular weight, and concentration of linear homopolymers of poly(methyl methacrylate) (PMMA) ranging from 12,470 to 65,700 g/mol Mw in a proper solvent. The researchers experimentally determined the critical chain overlap concentration, c*, the crossover concentration between the dilute and the semidilute concentration regimes, which was in good agreement with the theoretically determined value, estimated by the criteria $c^* \sim 1/[\eta]$ ($[\eta]$ is the intrinsic viscosity; c* is the critical chain overlap concentration), where the intrinsic viscosity was estimated from the Mark-Houwink parameters, K and a (at 25 8°C in DMF) obtained from the literature [20]. The experiment indicated that the value of the zero shear viscosity with the c/c* had a certain relationship with morphology of the fibers. As the concentration was increased (semidilute untangled regime $1 < c/c^* < 3$), polymer droplets and some bead-on-string fibers were observed. Upon further increase in concentration (semidilute entangled regime), bead-on-string fibers were obtained at $3 < c/c^* < 6$. Uniform fiber formation was observed at high concentration (semidilute entangled regime $c/c^* > 6$) [20]. Furthermore, it was found that the lower molecular weight polymer showed greater critical chain overlap concentration solution, increasing the solution concentration needed to form corresponding fibers described above [20]. Zhang et al. further explored the effect of various concentrations of PEO solutions on the morphology of bead-on-string fibers. The results showed that there were obvious changes in the morphology of bead-on-string fibers, showing increased fiber diameter and the main diameter of beads changed from 300–600 nm to 900–1500 nm, when concentration of PEO solution changed from 2.5 to 7.5% [67]. Overall, viscosity of polymer solution had an important effect on the formation of bead-on-string fibers, which can be inhibited by increasing the viscosity of the polymer solution.

3.3. Charge density of spinning solution

In general, the charge density of solution is usually characterized by density of charge distribution. Electrospinning is a process where the jet is stretched under an electric force. The greater charge density the solution has, the easier charge fluctuations take place, which makes jet bend in the axis of the electric field causing the occurrence of whipping instability, allowing jet to split into small branches to form nanoscale fibers [24]. When the fiber can be properly obtained, lower charge density polymer and smaller electric force jet makes harder the jet slender, making easier bead-on-string fibers fabrication. As the charge density of jets increase, the diameter of beads become smaller, making the morphology of the beads change from spherical to spindle. Geng et al. proved this theory [44]. The results showed that the morphology of fibers had significant change at different electric fields strengths. When the electric field strength was 1 kV/cm, the product contained spindle-shaped beads, presenting a crude fiber. As the electric field was strengthened to 3–4.5 kV/cm, uniform fibers were observed. However, when the values exceeded 4.5 kV/cm, many bead-on-string fibers were obtained. Fong et al. also studied the relationship between charge density and morphology of fiber [25]. The results indicated that the size of beads became smaller, the number of beads decreased, and the shape of beads changed from spherical to spindle with increasing charge density. When the charge density increased to a certain value, charge neutralization of jet due to corona discharge phenomenon occurs. It led to the decrease of charge density and then the jet cannot be drawn efficiently to form bead-on-string fibers. Therefore, the charge density of polymer solution is key to the formation of bead-on-string fibers without taking into account the surface tension and viscosity of the polymer solution.

The main influencing factors of charge density of polymer solution are the electrical conductivity of the polymer solution and the electrospinning parameters. Therefore, bead-on-string fibers can be obtained by adjusting electrical conductivity and electrospinning parameters.

In general, solution charge density is used to characterize the electrical conductivity which characterizes ability to conduct current. The greater electrical conductivity the solution has, the higher unit charge density the jet solution has, which allows better conductive solution and allows the jet to be drawn to form smooth electrospun fibers within certain range. Electrical conductivity can be changed by adding salt ions. Nartetamrongsutt et al. studied the effect of salt ions on the morphology of bead-on-string fibers through added salt ions in the same polyvinylpyrrolidone (PVP) solutions to change electrical conductivity of spinning. Data showed electrical conductivity became larger with the increase of salt ions in the PVP solution. The results indicated that the greater electrical conductivity and viscosity the polymer solution had, the easier Rayleigh instability was limited, which was favorable for the formation of uniform fiber [48]. On this foundation, Liu et al. added 0.5 wt% and 1.0 wt% LiCL solution in 14 wt% poly(butylenes succinate)/(chloroform/2-chloroethanol), respectively, and electrospun under same condition [41]. The results were consistent with Fong. This is because the electrical conductivity of the solution became larger by adding salt ions that can increase electrical density to draw jet easily under same voltage. The morphology of the beads changed from spherical to spindle at the same voltage [41].

The applied voltage is the most influential factor to the charge density of the jet in the electrospinning parameter. Only under the function of applied voltage the liquid drop at the end of the spraying device can be ejected to form fiber, when the electric field force is larger than the viscous resistance and the surface tension. The charge on the surface of jet and the electrical force jet suffered increases with increasing the applied voltage. This theory has been proved by many experiments [19]. A previous report by Jarusuwannapoom et al. [43] explored the effect of voltage on the number of beads with polystyrene (PS)/1,2-dichloroethane spinning solution at various voltages. The results showed that when the concentration of the spinning solution was 10 wt%, the number of beads increased as voltage changed from 15 to 25 kV. The reason is that low concentration and viscosity solutions led to lack of molecular chain entanglement, which caused jet to be drawn sharply to produce instability of jet with increasing electric density and repulsion between charges. Owing to low viscous resistance of polymer solution, viscous resistance cannot compete with surface tension, and therefore, jet split into smaller drops and shrunk into spheres, forming bead-on-string fibers. However, when concentration increased to 20 wt%, the number of beads decreased with increasing voltage. Under high concentration, the viscosity increased and the electric density of the jet also increased, increasing the repulsion between charges and allowing the jet to be drawn. Meanwhile, surface tension could not overcome the viscous resistance, making jet to break and shrink due to high viscosity. Therefore, the jet continuously moved in the form of cylinder to eventually form fibers [43].

3.4. Other factors

Collecting distance, spinning rate, solvent evaporation rate, environmental temperature and humidity also can affect the formation of beads. The collecting distance is characterized by the distance from the nozzle to the collecting plate. When the motion and movement time of the jet became longer and the time for solvent evaporation became longer in the electric field, the number of beads decreased and the diameter of beads became smaller. Meanwhile, the increase of collecting distance reduced electric field force to weaken stretch jet suffered, which was favorable to decrease number of beads. The final morphology of fiber depends on the contribution of both aspects [46]. Generally speaking, the faster speed at which the jet moves, shorter the residence time of the jet in air. On the one hand, molecule chains of polymer solution cannot draw enough under electric field force. On the other hand, there is no time to completely evaporate the solvent, therefore forming the bead-on-string fibers. Environmental parameters, including temperature humidity and air velocity, can also affect the formation of bead-on-string fibers. The effect of environmental factors on spinning is not easy to control and adjust. Therefore, keeping the external environmental factors stable and not interfering with the movement of jet process are also important [47, 68].

4. Characterization and morphology of the beads within bead-on-string fibers

To characterize the bead-on-string fibers, normally electron microscopes, optical microscopes are used [18, 26, 44]. Through the images obtained from those microscopes, it can be seen that the morphology of beads within the bead-on-string fibers were not always the same. Generally, spherical- and spindle-shaped beads are taken the most part. However, there are also some heteromorphic ones, such as concave beads [19]. According to the formation conditions of bead-on-string fibers, the morphology of beads can be tuned by changing various processing parameters, such as polymer concentration, which can determine the viscosity of the electrospinning solution, applied voltage and solution electrical conductivity, which are related to the charge density of the solutions, and collecting distance [19, 48]. An increase in the viscosity of the solution leads to a less likely formation of the beads. Higher charge density decreases the possibility of beads formation. Longer collecting distance was favorable to decrease numbers and diameters of beads. According to our recent study, the silk fibroin/PEO bead morphology and aspect ratio can be tuned by changing solution properties, applied voltage, and collecting distances. Our results are consistent with the previous studies [19, 48].

5. Applications of electrospun bead-on-string fibers

5.1. Drug delivery

It is generally believed that the smaller size of the capsules, the larger surface area the capsule and drug occupy, presenting a faster decomposition rate of the drug and making easier the drug to be absorbed by the body. Electrospun fibers have many outstanding advantages as drug carriers due to their nanoscale, which allow drugs that are generally difficult to be absorbed by the body to present a slow delivery and hence enhance the absorption and produce better treatment effect. Furthermore, biodegradable materials that are used as drug carriers can be broken down into small molecules that are absorbed or expelled from the body to accomplish treating function with the release of the drug [68]. However, there are some shortcomings, such as for instance, the drugs often found on the fiber surface and cannot be coated completely causing drugs to be partially exposed, which can lead to burst release [14, 69, 70]. Therefore, the achievement of sustained release must overcome the problem of drugs' burst release.

The characteristics of the alternating distribution of sub-nanofiber and sub-micron beads of the electrospun bead-on-string fibers meet the demand of drug loading and sustained release. Bigger beads not only can complete coat water-soluble drugs or solid particles, but also can realize the function of easy absorption and fast decomposition of material to achieve drug sustained-release. Many researchers have studied the function of electrospun bead-on-string fibers in drug sustained release. The initial report was performed by Qi et al. that first dissolved AOT (sodium bis(2-ethylhexyl)sulfosuccinate) in dichloromethane, where BSA was then added after crosslinking, and then fiber-forming materials poly(L-lactic acid) (PLLA, Mw = 500,000) was added to form water in oil emulsion, with which bead-on-string fibers loading BSA was successfully prepared [26]. The experiment compared the release effect within 120 h of smooth electrospinning fibers and bead-on-string fibers with the same amount of drug loading. The results showed that the smooth nanofibers showed a drug release rate of up to 80% in 10 h and the phenomenon of burst release of drug was significant. However, when the drug was loaded on the bead-on-string fibers, the release amount of drug was below 50%, which indicated that the defect, burst release of drug of smooth fiber as drug carrier, can be improved by bead-on-string fibers to realize sustained release and control of drug. He further studied the storage locations of BSA in bead-on-string fibers by the vapor etching. It was found that most drugs were coated on the beads. Somvipart et al. also proved that bead-on-string fibers can be used to achieve sustained drug release [42]. In the experiment, methylene blue was added into the spinning solution to prepare both smooth and bead-on-string fibers. The test of drug delivery *in vitro* showed that the released amount of methylene blue from smooth fiber reached 80%; however, the release amount of methylene blue from bead-on-string fibers was only 50% which achieved sustained drug release. To summarize, bead-on-string fibers can be used as drug carrier to achieve sustained drug release.

5.2. Filter material

With the rapid development of the society, environmental pollution leads to poor air quality which results in frequent occurrences of respiratory diseases and lung diseases that bring great harm to people's life and health. Therefore, the development of high-efficiency air-filtrating materials has become a pressing requirement. According to previous studies, traditional nonwoven fiber materials have been widely used in the field of air filtration, such as glass fiber or melt-blown fiber, which have been used as indoor air filter or the core filter media of N95 respirators. However, the filtration efficiency is relatively low for fine particles since the micron-sized pore of the material are not small enough [71].

The electrospinning nanofibers that are characterized by small size, large specific surface area, high porosity and good connectivity are considered as the most promising materials for air filtration media [72, 73]. However, electrospinning nanofibers used as air filter material have higher packing density, hindering the flow of air [74, 75]. Porous bead-on-string fibers that have high specific surface area and porosity, and beads covered with micropores not only can improve packing density but also benefit air flow and adsorption for the solid particles in the air [76]. Yun et al. studied the filtration properties of the fiber mat with different morphologies by preparing bead-on-string fiber mat and particle/nanofiber composite. The results showed that the quality factor of the bead-on-string fiber mat and composite fiber mat was higher than nanofiber mats [76]. Wang et al. studied the filtration of bead-on-string fibers and found that the existence of beads could greatly improve the efficiency of air filtration. In this experiment, the bead-on-string fibers formed from 5 wt% poly(lactic acid) solution and solvent mixture containing dichloromethane (DCM)/N,N-dimethylacetamide (the ratio is 1:10) exhibited excellent filtration efficiency (99.997%) and a low pressure drop (165.3 Pa), which are promising characteristics for the membranes' application as filters for respiratory protection, indoor air purification, and other filtration applications [45].

6. Conclusions

Electrospun bead-on-string fibers have been successfully prepared by adjusting the concentration, surface tension, and charge density of the spinning solutions and electrostatic spinning parameters. Bead-on-string fibers characterized by alternating distribution of sub-nano fiber and sub-micron beads have shown great potential for the application of sustained drug release and air filtration systems. Recently, with the increase of the applications of bead-on-string fibers, much research has been performed on the formation mechanism of bead-on-string fibers. It is hoped that morphology and number of bead-on-string fibers can be controlled accurately to achieve particular applications. These research results have great significance for drug-sustained release, administration of air pollution, and improvement of water quality.

The mechanism of bead-on-string fibers formation has been proved to be caused by axisymmetric instability of jet. However, there has been few theoretical and fundamental research and no systematic summary has been prepared. Furthermore, the effective formulas and models to theoretically study and predict bead-on-string fibers have not yet been established. Therefore, the establishment of systematic theoretical, effective models, and formulas of bead-on-string fibers is essential for its research and applications.

Acknowledgements

This work was supported by Natural Science Foundation of Jiangsu Province (No. BK20161254, BK2012634) and National Natural Science Foundation of China (No. 51103092).

Author details

Huijing Zhao* and Huanjie Chi

*Address all correspondence to: zhhj@suda.edu.cn

National Engineering Laboratory for Modern Silk, College of Textile and Clothing Engineering, Soochow University, Suzhou, China

References

[1] Greine A, Wendorff J. Electrospinning: A fascinating method for the preparation of ultrathin fibers. Angewandte Chemie International Edition. 2007;**46**:5670-5703

[2] Hu X, Liu S, Zhou G, Huang Y, Xie Z, Jing X. Electrospinning of polymeric nanofibers for drug delivery applications. Journal of Controlled Release. 2014;**185**:12-21

[3] Chakraborty S, Liao I, Adler A, Leong K. Electrohydrodynamics: A facile technique to fabricate drug delivery systems. Advanced Drug Delivery Reviews. 2009;**61**:1043-1054

[4] Binulal N, Natarajan A, Menon D, Bhaskaran V, Mony U, Nair S. PCL–gelatin composite nanofibers electrospun using diluted acetic acid–ethyl acetate solvent system for stem cell-based bone tissue engineering. Journal of Biomaterials Science. Polymer Edition. 2014;**25**:325-340

[5] Schmedake T, Cunin F, Link J, Sailor M. Standoff detection of chemicals using porous silicon "smart dust" particles. Advanced Materials. 2002;**14**:1270-1272

[6] Martindale D. Medicine - Tissue repair - Scar no more. Scientific American. 2000;**283**:34-36

[7] Furmhals A. 1934. US patent 1975504

[8] Simm W, Gosline C, Bonart R, Von Falkai B. 1979. US 4143196 A

[9] Taylor G. Disintegration of water drops in electric field. In: Proceedings of the Royal Society of London. Series A, Mathematical and Physical Sciences. 1964;**280**:383-397

[10] Baumgarten P. Electrostatic Spinning of acrylic microfibers. Journal of Colloid and Interface Science. 1971;**36**:71-79

[11] Doshi J, Reneker D. Electrospinning process and applications of electrospun fibers. Journal of electrostatics. 1995;**35**:151-160

[12] Subbiah T, Bhat G, Tock R, Parameswaran S, Ramkumar S. Electrospinning of nanofibers. Journal of Applied Polymer Science. 2005;**96**:557-569

[13] Langer R. Biomaterials in drug delivery and tissue engineering: One laboratory's experience. Accounts of Chemical Research. 2000;**33**:94-101

[14] Tian L, Prabhakaran M, Ding X, Kai D, Ramakrishna S. Emulsion electrospun vascular endothelial growth factor encapsulated poly(l-lactic acid-co-ε-caprolactone) nanofibers for sustained release in cardiac tissue engineering. Journal of Materials Science. 2012;**47**:3272-3281

[15] Frokjaer S, Otzen D. Protein drug stability-a formulation challenge. Nature Reviews Drug Discovery. 2005;**4**:298-306

[16] Kaur S, Sundarrajan S, Rana D, Matsuura T, Ramakrishna S. Influence of electrospun fiber size on the separation efficiency of thin film nanofiltration composite membrane. Journal of Membrane Science. 2012;**392**:101-111

[17] Shin C, Chase G, Reneker D. The Effect of Nanofibers on Liquid – Liquid Coalescence Filter Performance. AICHE Journal. 2005;**51**:3109-3113

[18] Shin C. Filtration application from recycled expanded polystyrene. Journal of Colloid and Interface Science. 2006;**302**:267-271

[19] Lee K, Kim H, Bang H, Jung Y, Lee S. The change of bead morphology formed on electrospun polystyrene fibers. Polymer. 2003;**44**:4029-4034

[20] Gupta P, Elkins C, Long TE, Wilkes G. Electrospinning of linear homopolymers of poly(methyl methacrylate): Exploring relationships between fiber formation, viscosity, molecular weight and concentration in a good solvent. Polymer. 2005;**46**:4799-4810

[21] Yarin A, Zussman E. Upward needleless electrospinning of multiple nanofibers. Polymer. 2004;**45**:2977-2980

[22] Liu Y, He J. Bubble electrospinning for mass production of nanofibers. International Journal of Nonlinear Sciences and Numerical Simulation. 2007;**8**:393-396

[23] Varesano A, Carletto R, Mazzuchetti G. Experimental investigations on the multi-jet electrospinning process. Journal of Materials Processing Technology. 2009;**209**:5178-5185

[24] Reneker D, Yarin A. Electrospinning jets and polymer nanofibers. Polymer. 2008;**49**: 2387-2425

[25] Fong H, Chun I, Reneker D. Beaded nanofibers formed during electrospinning. Polymer. 1999;**40**:4585-4592

[26] Qi H, Hu P, Xu J, Wang A. Encapsulation of drug reservoirs in fibers by emulsion electrospinning: Morphology characterization and preliminary release assessment. Biomacromolecules. 2006;**7**:2327-2330

[27] Carrillo C, Nypelo T, Rojas O. Double emulsions for the compatibilization of hydrophilic nanocellulose with non-polar polymers and validation in the synthesis of composite fibers. Soft Matter. 2016;**12**:2721-2728

[28] Sanders E, Kloefkorn R, Bowlin G, Simpson D, Wnek G. Two-phase electrospinning from a single electrified jet: Microencapsulation of aqueous reservoirs in poly(ethylene-co-vinyl acetate) fibers. Macromolecules. 2003;**36**:3803-3805

[29] Xu X, Yang L, Xu X, Wang X, Chen X, Liang Q, Zeng J, Jing X. Ultrafine medicated fibers electrospun from W/O emulsions. Journal of Controlled Release. 2005;**108**:33-42

[30] Angeles M, Cheng H, Velankar S. Emulsion electrospinning: Composite fibers from drop breakup during electrospinning. Polymers for Advanced Technologies. 2008;**19**:728-733

[31] Xu X, Zhuang X, Chen X, Wang X, Yang L, Jing X. Preparation of core-sheath composite nanofibers by emulsion electrospinning. Macromolecular Rapid Communications. 2006;**27**:1637-1642

[32] Li D, Xia Y. Preparation of core-sheath composite nanofibers by emulsion electrospinning. Nano Letters. 2004;**4**:933-938

[33] Loscertales I, Barrero A, Guerrero I, Cortijo R, Marquez M, Ganan-Calvo A. Micro/nano-encapsutation via electrified coaxial liquid jets. Science. 2002;**295**:1695-1698

[34] Sun Z, Zussman E, Yarin A, Wendorff J, Greiner A. Compound core–shell polymer nanofibers by co-electrospinning. Advanced Materials. 2003;**15**:1929-1932

[35] Larsen G, Spretz R, Velarde-Ortiz R. Use of coaxial gas jackets to stabilize taylor cones of volatile solutions and to induce particle-to-fiber transitions. Advanced Materials. 2004;**16**:166-169

[36] Zhang Y, Wang X, Feng Y, Li J, Lim C, Ramakrishna S. Coaxial electrospinning of (fluorescein isothiocyanate-conjugated bovine serum albumin) encapsulated poly(ε-caprolactone) nanofibers for sustained release. Biomacromolecules. 2006;**7**:1049-1057

[37] Tian X, Bai H, Zheng Y, Jiang L. Bio-inspired heterostructured bead-on-string fibers that respond to environmental wetting. Advanced Functional Materials. 2011;**21**:1398-1402

[38] Zuo W, Zhu M, Yang W, Yu H, Chen Y, Zhang Y. Experimental study on relationship between jet instability and formation of beaded fibers during electrospinning. Polymer Engineering & Science. 2005;**45**:704-709

[39] Yoon Y, Moon H, Lyoo W, Lee T, Park W. Superhydrophobicity of PHBV fibrous surface with bead-on-string structure. Journal of Colloid and Interface Science. 2008;**320**:91-95

[40] Zhao L, Song C, Zhang M, Zheng Y. Bioinspired heterostructured bead-on-string fibers via controlling the wet-assembly of nanoparticles. Chemical Communications. 2014;**50**:10651-10654

[41] Liu Y, He J, Yu J, Zeng H. Controlling numbers and sizes of beads in electrospun nanofibers. Polymer International. 2008;**57**:632-636

[42] Somvipart S, Kanokpanont S, Rangkupan R, Ratanavaraporn J, Damrongsakkul S. Development of electrospun beaded fibers from Thai silk fibroin and gelatin for controlled release application. International Journal of Biological Macromolecules. 2013;**55**:176-184

[43] Jarusuwannapoom T, Hongrojjanawiwat W, Jitjaicham S, Wannatong L, Nithitanakul M, Pattamaprom C, Koombhongse P, Rangkupan R, Supaphol P. Effect of solvents on electro-spinnability of polystyrene solutions and morphological appearance of resulting electrospun polystyrene fibers. European Polymer Journal. 2005;**41**:409-421

[44] Geng XY, Kwon O, Jang J. Electrospinning of chitosan dissolved in concentrated acetic acid solution. Biomaterials. 2005;**26**:5427-5432

[45] Wang Z, Zhao C, Pan Z. Porous bead-on-string poly(lactic acid) fibrous membranes for air filtration. Journal of Colloid and Interface Science. 2015;**441**:121-129

[46] Buchko C, Chen L, Shen Y, Martin D. Processing and microstructural characterization of porous biocompatible protein polymer thin films. Polymer. 1999;**40**:7397-7407

[47] Zhang C, Yuan X, Wu L, Han Y, Sheng J. Study on morphology of electrospun poly(vinyl alcohol) mats. European Polymer Journal. 2005;**41**:423-432

[48] Nartetamrongsutt K, Chase G. The influence of salt and solvent concentrations on electrospunpolyvinylpyrrolidone fiber diameters and bead formation. Polymer. 2013;**54**:2166-2173

[49] Taylor G. Electrically driven jets. Proceedings of the Royal Society of London. Series A, Mathematical and Physical Sciences. 1969;**313**:453-475

[50] Cloupeau M, Prunet-Foch B. Electrostatic spraying of liquids in cone-jet mode. Journal of Electrostatics. 1989;**22**:135-159

[51] Hohman M, Shin M, Rutledge G, Brenner M. Electrospinning and electrically forced jets. I. Stability theory. Physics of Fluids. 2001;**13**:2201-2220

[52] Yu J, Fridrikh S, Rutledge G. The role of elasticity in the formation of electrospun fibers. Polymer. 2006;**47**:4789-4797

[53] Yarin A, Koombhongse S, Reneker D. Bending instability in electrospinning of nanofibers. Journal of Applied Physics. 2001;**89**:3018-3026

[54] Reneker D, Yarin A, Fong H, Koombhongse S. Bending instability of electrically charged liquid jets of polymer solutions in electrospinning. Journal of Applied Physics. 2000;**87**:4531-4547

[55] Reneker D, Yarin A, Zussman E, Xu H. Electrospinning of nanofibers from polymer solutions and melts. Advances in Applied Mechanics. 2007;**41**:43-346

[56] Hohman M, Shin M, Rutledge G, Brenner M. Electrospinning and electrically forced jets. II. Applications. Physics of Fluids. 2001;**13**:2221-2236

[57] Huebner A, Chu H. Instability and breakup of charged liquid jet. Journal of Fluid Mechanics. 1971;**49**:361-372

[58] Shin Y, Hohman M, Brenner M, Rutledge G. Experimental characterization of electrospinning: the electrically forced jet and instabilities. Polymer. 2001;**42**:09955-09967

[59] Shin Y, Hohman M, Brenner M, Rutledge G. Electrospinning: A whipping fluid jet generates submicron polymer fibers. Applied Physics Letters. 2001;**78**:1149-1151

[60] Lezzi A, Prosperetti A. Rayleigh-taylor instability for adiabatically stratified fluids. Physics of Fluids A. 1989;**1**:1784-1795

[61] Christanti Y, Walker L. Surface tension driven jet break up of strain-hardening polymer solutions. Journal of Non-Newtonian Fluid Mechanics. 2001;**100**:9-26

[62] Magarvey R, Outhouse L. Note on the break-up of a charged liquid. Journal of Fluid Mechanics. 1962;**13**:151-157

[63] Zhang C, Yu S. Nanoparticles meet electrospinning: Recent advances and future prospects. Chemical Society Reviews. 2014;**43**:4423-4448

[64] Cui W, Li X, Zhou S, Weng J. Investigation on process parameters of electrospinning system through orthogonal experimental design. Journal of Applied Polymer Science. 2007;**103**:3105-3112

[65] Ku B, Sang S. Electrospray characteristics of highly viscous liquids. Journal of Aerosol Science. 2002;**33**:1361-1378

[66] Ki C, Baek D, Gang K, Lee K, Um I, Park Y. Characterization of gelatin nanofiber prepared from gelatin–formic acid solution. Polymer. 2005;**46**:5094-5102

[67] Zhang Y, Li T, Ding X, Hu J, Yang X. Effects of polyethylene oxide concentration on the size of beads in electrospun beaded nanofibers. Journal of Donghua University (English Edition). 2014;**31**:511-514

[68] Gentsch R, Boysen B, Lankenau A, Borner H. Single-step electrospinning of bimodal fiber meshes for ease of cellular infiltration. Macromolecular Rapid Communications. 2010;**31**:59-64

[69] George M, Abraham T. Polyionic hydrocolloids for the intestinal delivery of protein drugs: Alginate and chitosan - A review. Journal of Controlled Release. 2006;**114**:1-14

[70] Wu F, Jin T. Polymer-based sustained-release dosage forms for protein drugs, challenges, and recent advances. AAPS PharmSciTech. 2008;**9**:1218-1229

[71] Hung C, Leung W. Filtration of nano-aerosol using nanofiber filter under low Peclet number and transitional flow regime. Separation and Purification Technology. 2011;**79**:34-42

[72] Barhate R, Ramakrishna S. Filtration problems and solutions from tiny materials. Journal of Membrane Science. 2007;**296**:1-8

[73] Desai K, Kit K, Li J, Davidson P, Zivanovic S, Meyer H. Filtration problems and solutions from tiny materials. Polymer. 2009;**50**:3661-3669

[74] Yun K, Suryamas A, Iskandar F. Morphology optimization of polymer nanofiber for applications in aerosol particle filtration. Separation and Purification Technology. 2010;**75**:340-345

[75] Mikheev A, Kanev I, Morozova T. Water-soluble filters from ultra-thinpolyvinylpirrolidone nanofibers. Journal of Membrane Science. 2013;**448**:151-159

[76] Yun K, Hogan C, Matsubayashi Y, Kawabe M, Iskandar F, Okuyama K. Nanoparticle filtration by electrospun polymer fibers. Chemical Engineering Science. 2007;**62**:4751-4759

Chapter 6

Electrospinning of Functional Nanofibers for Regenerative Medicine: From Bench to Commercial Scale

Chris J. Mortimer, Jonathan P. Widdowson and
Chris J. Wright

Additional information is available at the end of the chapter

http://dx.doi.org/10.5772/intechopen.73677

Abstract

Nanofibers are an important material for regenerative medicine as they have a commensurate morphology to that of the macromolecular matrix that supports and houses the growth of cells and tissues within the body. Electrospinning is widely used to fabricate non-woven structures on the nanoscale and the versatility of the technique has widened the application of nanofibers. This is due to ease of extending nanofiber functionality through the incorporation of active materials both during and after electrospinning. Recent developments in electrospinning devices, such as needle-free systems, have reinvigorated research as these advances now allow fabrication of nanofibers at commercial scales. The process of electrospinning has a number of operating parameters that are adjusted in optimisation to achieve ideal fibres and a multitude of instrument configurations can be adopted to achieve the required manufacture. The innate properties of nanofibers, such as high surface area to volume ratio, have many proven benefits for regenerative medicine and the chapter examines these before discussing how functionality can be further improved. Numerous materials can be incorporated in the manufacture of electrospun mats, however when choosing materials for regenerative medicine, biocompatibility and biodegradability are the dominant functionalities that are required.

Keywords: electrospinning, tissue engineering, wound healing nanofibers, biomaterials, antimicrobials

IntechOpen

© 2018 The Author(s). Licensee IntechOpen. This chapter is distributed under the terms of the Creative Commons Attribution License (http://creativecommons.org/licenses/by/3.0), which permits unrestricted use, distribution, and reproduction in any medium, provided the original work is properly cited. (cc) BY

1. Introduction

Electrospinning is an extremely versatile technique for the production of nanofibers. As a consequence, electrospun fibres have been fabricated for a wide range of applications from separation processes to tissue engineering. The versatility of the electrospinning process has allowed the functionality of the nanofibers to be extended beyond the innate improvement of properties enabled by the fabrication of materials with nanoscale dimensions. Further functionality has been achieved by the incorporation of nanoparticles and other bioactive compounds, this has been particularly important for the application of nanofibers for tissue engineering, wound healing and drug delivery; the three themes of regenerative medicine. The developments of regenerative medicine that we seem to be witnessing every day is just one example of the increasing demand not just for novel nanofiber constructs but manufacture of functional nanofibers at economic scales, whether that is at high value and low volume, as in tissue engineering scaffolds or high volume manufacture as in wound dressings. Indeed, scale of manufacture is another advantage for the application of electrospinning, as relatively recent instrument developments have reinvigorated the research area through the increase in volume of manufacture that they now allow. Thus, this chapter will examine the state of the art technology for electrospinning in the context of improved functionality and scale of manufacture, which is essential for the modern healthcare system and the realisation of the potential that regenerative medicine promises.

2. Electrospinning

2.1. Electrospinning process

Electrospinning uses a high voltage power supply to create a large potential difference between a grounded "collector" structure and a polymer solution or melt being delivered at a constant rate through an aperture, such as a blunt end needle. As the voltage is increased the like charges within the polymer fluid directly oppose surface tension, resulting in the normally spherical droplet at the aperture distending into a conical shape. This cone is referred to as the "Taylor" cone, after Sir Geoffrey Taylor who first mathematically modelled the phenomenon [1–3]. At a critical voltage the electrostatic attractive force between the solution and the collector causes a jet of polymer solution to be expelled from the cone tip towards the grounded collector surface. This jet then undergoes a whipping instability and dries in flight, depositing the nanofibers on the collector [4] (**Figure 1**).

It is the interaction of the applied electrical field and the electrical charge which is being carried by the jet which generates the tensile force required for electrospinning [5]. A stable electrospinning jet can be described as having four distinct regions [5]. These four regions are the base, the jet, the splaying and the collection regions. The base region can be referred to as the "Taylor Cone", where the jet is expelled from the polymer solution [1, 2]. This is where the polymer becomes electrically charged. When the voltage is increased to a critical voltage the

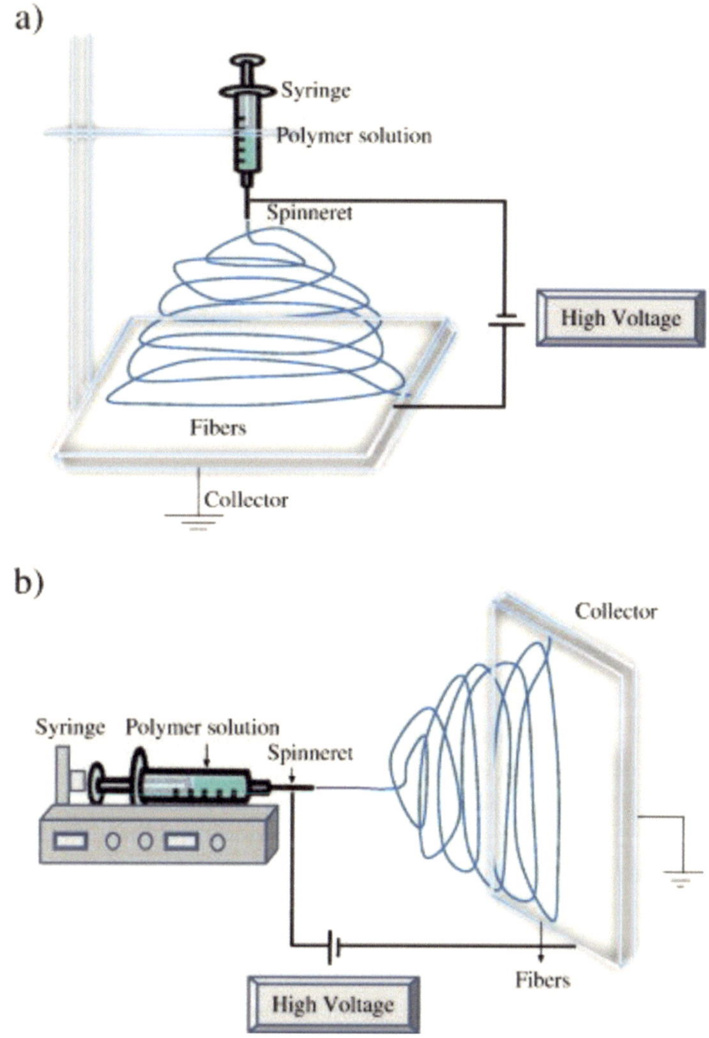

Figure 1. A schematic diagram of electrospinning apparatus in (a) a vertical setup and (b) a horizontal setup. Reprinted from Copyright (2010), with permission from Elsevier [4].

electrostatic attractive force between the solution and the collector cause a jet of polymer solution to be expelled from the cone as the repulsive forces generated by the applied electric field overcome the surface tension of the polymer. Once the jet is expelled from the Taylor cone it continues to accelerate towards the grounded collector. This region can be described as the jet region. The splaying region is where the jet undergoes a whipping instability. At the tip of the spinneret where the Taylor cone forms the jet is very stable but as it accelerates towards the collector and the solvent evaporates, the jet then undergoes what can be described as a whipping instability. This whipping instability is caused by an electrostatic repulsion within the polymer solution which is initiated as small bends in the fibres. It is the combination of acceleration of the jet and also evaporation of the solution which causes the jet to stretch decreasing its diameter. As this occurs the electric charge on the jet causes it to stretch in the radial direction, which as the jet becomes smaller allows the jet to split into two or more fibres as the radial forces manage to overcome the forces that are holding the jet together. This continues to occur as the jet travels towards the collector. Multiple fibres are formed all with like charges which then repel each other resulting in the 'splaying' effect causing a number of fibres to be deposited on the collector.

Despite the relative simplicity of the equipment involved, by carefully controlling processing parameters the fibre's diameters, orientation, total mat porosity and other properties can be controlled, allowing optimisation of the mat for a given application. In addition, the technique's ability to work with a wide variety of materials allows a range of specific biological, mechanical or chemical properties to be achieved [6]. Therefore by controlling solution properties such as the viscosity, conductivity, molecular weight and surface tension along with processing parameters such as the applied electrical field, distance from the syringe tip to the collector and flow rate of the polymer solution, a range of desirable characteristics can be attributed to the nanofibers [4].

2.2. Controllable variables and their effects

2.2.1. Solution properties

An increase in polymer concentration in a given solution will result in an increase in the solutions viscosity. It has been shown by a number of research groups that a decrease in polymer concentration results in a decreased fibre diameter [7–10]. Although this offers a certain level of control as the polymer concentration decreases it also leads to beaded fibres and a wider size distribution. Mit-uppatham et al. demonstrated this with Polyamide-6 showing at low concentrations a large number of droplets. As solution concentration increased the number of beads decreased and fibre morphology improved [8]. Katti et al. also demonstrated that at lower concentrations when beading is present fibres deposited are wet and therefore tip to collector distance must be increased [7]. Electrospinning will only occur when the polymer concentration is high enough to allow adequate chain entanglement resulting in continuous formation of uniform nanofibers when a high enough voltage is supplied [11]. Rayleigh instability occurs when electrostatic repulsion of charges in the electrospinning jet tends to increase its surface area while the surface tension opposes this force reducing the surface area of the jet [12]. At polymer concentrations too low to initiate electrospinning no polymer chain entanglements can occur and as such the polymer solution cannot resist the Rayleigh instability sufficiently, which results in the break-up of the jet into droplets. Increasing charge density on the surface of the droplet leads to the formation of smaller droplets and results in electrospraying. At higher concentrations the high viscosity hinders jet stretching in the whipping region. The high viscosity can also lead to practical difficulties when pumping the solution through an aperture [11]. The viscosity is the governing parameter when changing the polymer concentration but changing the solution composition also has an effect on fibre diameter. As the polymer concentration increases, the proportion of polymer to solvent in the jet increases and as such fibre diameter increases. This is due to the higher volume of polymer remaining once the solvent has evaporated off.

The conductivity of the solution is a crucial parameter. When the conductivity is low this results in the production of fibres with a greater diameter. This is most likely due to poor jet elongation [10]. As the conductivity increases the fibre diameter tends to decrease. This shows a clear relationship between the conductivity of a solution and the level of jet elongation. As the conductivity of the solution is increased the jet undergoes a higher degree of elongation along its axis due to the repulsion of charges under the electric field [13]. Tan et al. reported fibres with beads as a result of a low conductivity which is again most likely due to insufficient jet elongation resulting in the solution 'spitting' [10]. Increasing the conductivity of a polymer solution can also initiate the electrospinning of smooth fibres at lower polymer concentrations [14]. This is because the increase in conductivity results in an increased charge density at the surface of the jets which decreases the likelihood of bead formation and improves fibre extension in the whipping region [11]. The conductivity of a polymer solution can be adjusted by the addition of an inorganic salt or ionic organic compound [11]. This addition can also affect the surface tension and dielectric constant of the solution which can make assessing the effect of conductivity difficult [11].

Viscosity of a polymer solution will increase with increased molecular weight, meaning a lower concentration of polymer is required to form non beaded fibres with tight size distributions. At higher molecular weights there are a greater number of chain entanglements and therefore a higher viscosity at equivalent concentration compared to a lower molecular weight. As a result even at a low polymer concentration a high molecular weight polymer can provide sufficient chain entanglements to overcome the effects of surface tension and result in a uniform jet [10].

There is no evidence to show that the surface tension of a solution affects the morphology of fibres although this does not mean the surface tension of the solution is irrelevant as if it becomes too high it can result in jet instability which can have a drastic effect on the electrospinning process. Surface tension can also be adjusted to induce beads formation [11]. Generally, it is the surface tension and solution viscosity that are used to determine the range of polymer-solvent combinations that electrospinning is possible. The spinning voltage at which electrospinning is initiated tends to increase with surface tension. Fridrikh et al. suggest that when all other parameters are kept constant a lower surface tension is desirable [15]. Unfortunately, this is not simple as surface tension varies both with varying concentrations and due to the chemical nature of the polymer [16]. Surface tension can be adjusted through the selection of different solvents or through the addition of a surfactant. A surfactant is a substance containing hydrophobic groups (head) and hydrophilic groups (tails) and they tend to reduce the surface tension between liquids and solids.

A solvent with a high vapour pressure at normal temperatures can be referred to as being volatile. During the electrospinning process this results in the jet spending less time in the elongation stage undergoing stretching as the solvent evaporates off. Therefore, a more volatile solvent (higher vapour pressure) will result in fibres of a larger average diameter. This allows the researcher to control the fibre diameter to a certain degree through the choice of solvent based on its vapour pressure, although a solvent with a particularly low vapour pressure may be unsuitable and result in the deposition of wet fibres, with coalescence at fibre junctions.

2.2.2. Processing parameters

It has been shown by a number of research groups that the applied voltage does not have a great effect on the fibre diameter [7, 10, 17]. Electrospinning will only occur at a certain range of voltage although if the voltage is too high this can result in multiple jets forming which can decrease diameter of fibres but widen the size distribution [10]. Generally, fibre diameter decreases as applied voltage increases due to the greater columbic forces causing greater stretching of the solution and resulting in a smaller fibre diameter with fibres drying more quickly [4, 9]. Baumgarten showed that as voltage increases jet diameter initially decreases and then increases as the voltage continues to increase [18]. As a result of this flow rate must increase as the voltages increases. It has also been reported that deposition rates increase with increased voltage [17]. This could explain the inconsistent reports of the effect that voltage has on fibre diameter and other parameters such as the feed rate and tip to collector distance must also be considered when observing the effects of applied voltage [11]. By measuring the current flow Deitzel et al. showed that for a PEO/water system the increase in mass flow rate is almost linear with applied voltage [16]. In increase in voltage which results in an increased mass flow rate can also result in spinning from within the needle as the Taylor cone recedes which can result in uneven and beaded fibres [19].

The flow rate is the rate at which the solution is pumped through the needle to maintain a droplet of solution available for Taylor cone formation. The flow rate of the solution has not been shown to have a significant effect on the fibre diameter or morphology [10] although some studies have shown an increase in flow rate can result in an increase in fibre diameter [20]. The ideal flow rate should match the rate at which the solution is being ejected from the tip [11]. At lower flow rates electrospinning tends to be intermittent, with the Taylor cone often receding into the needle as previously mentioned.

If the distance between the spinneret and the collector is either too small or too large, the result is beaded fibres. The ideal distance is the minimum distance required to allow the fibres enough time to dry between the spinneret and the collector [17, 18]. Tip to collector distance is usually selected according to the vapour pressure of the solvent. When increasing the distance, the voltage applied to the spinneret needs to be increased to maintain the electric field. Further to these controllable variables it has been shown that by modifying the collector architecture, for example by using two parallel conductive substrates of varying gap size, fibres can be aligned uniaxially into arrays [21]. This alignment of fibres has led to anisotropic mat properties in terms of tensile strength as well as directional cell growth, as previously discussed. Further modifications to the collector have been illustrated to expand the possible fibre orientations including rotating drum electrode [22] and knife-edge collectors [23]. Furthermore, "coaxial" electrospinning allows for a more complex fibre architecture, forming a fibre comprised of two non-mixed materials in a core-sheath arrangement [24].

2.2.3. Collector modifications

To produce aligned fibres, one of the techniques used employs a rotating drum electrode as a collector. In this setup the standard collector (usually flat, aluminium foil or similar) is replaced by a rotating mandrel [22]. When the rotational speed of the mandrel is sufficiently high it results in electrospun fibres which are aligned along the rotational axis of the mandrel.

Teo et al. described a knife-edged collector system [25] adapted from a similar setup described by Bornat [26]. The setup consisted of a parallel grid with a negative charge applied to a knife-edged aluminium bar. A Teflon tub was allowed to rotate between the electrospinning jet and knife-edged aluminium bar while fibres were electrospun on to the Teflon tube [25]. This technique was also used to electrospin tubular structures aligned in a diagonal direction.

A similar setup to the rotating drum electrode is the rotating disk electrode. In this setup instead of electrospinning onto a rotating mandrel a rotating disk is used as a collector [27]. This results in fibres which are aligned relatively to the disk.

Li et al. demonstrated the ability of uniaxially aligned arrays of nanofibers formed using parallel electrodes as collectors [21]. Their setup consisted of a standard system where a needle was aligned vertically electrospinning downwards where the collector consisted of two strips of conductive silicon separated by a gap. As the fibres are formed the ends of the fibres (those which are closest to the collector strips) will generate a stronger electrical force than anywhere else in the fibre. This results in the fibre being stretched across the gap and alignment along a single axis (**Figure 2**).

2.2.3.1. Control of the electric field

The path of an electrospinning jet can be altered by an electric field using its charge [28]. This can be achieved through the addition of auxiliary electrodes which can be used to align nanofibers (using two collectors separated by a gap), form simple patterns and control the deposition of nanofibers [21, 22, 29]. The auxiliary electrodes could include a base electrode, steering electrodes, focussing electrodes, guiding electrodes and the collector itself [28]. Yang et al. showed the use of a base electrode results in a more uniform electric field meaning a larger voltage can be applied to the nozzle, increasing the average field strength and resulting in fibres of a smaller diameter [30]. This could also be seen as a disadvantage as the use of a base electron means a higher voltage is required. A focusing electrode can be used to 'dampen' the chaotic motion of the electrospinning jet so the fibres are deposited in a more localised area. These are usually shaped as ring surrounding the jet but can also be a cylinder or cone. Using multiple pairs of steering electrodes can allow complex patterns to be fabricated.

2.2.3.2. Coaxial electrospinning

One method, coaxial electrospinning allows the encapsulation of materials which cannot usually be electrospun in to a core-shell fibre. A basic coaxial electrospinning setup will consist of a coaxial needle with 2 separate syringe pumps, one pumping a solution (conductive or non-conductive) through the inner lumen of the needle and another pumping a conductive electrospinnable solution through the outer needle. By adjusting the respective flow rates and applying a suitable electric current the solution being pumped through the outer lumen can be electrospun forming fibres encapsulating the solution being pumped through the inner lumen [31, 32]. This technique makes electrospinning a suitable fabrication method for a number of applications including for the controlled release of drugs and proteins [33], for nanowires

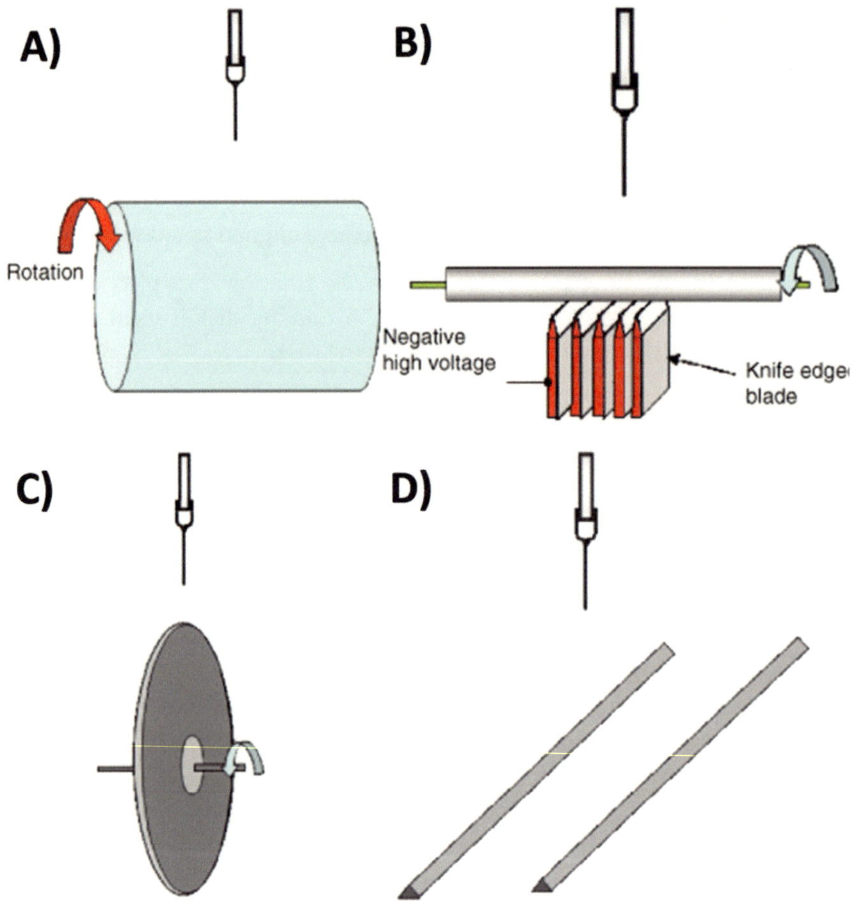

Figure 2. A schematic diagram of electrospinning apparatus in (a) a vertical setup and (b) a horizontal setup. Reprinted from Copyright (2010), with permission from Elsevier [4].

offering a number of potential applications in microelectronics and optical electronics [34] and also for the incorporation of stem cells in to tissue engineering scaffolds through a technique such as cell electrospinning [35, 36]. The latter of these applications is discussed below.

Once a polymer scaffold has been developed it can then be manually seeded with cells, frequently the cells employed for this are stem cells. The scaffolds are usually seeded with cells by adding a concentrated cell suspension to a suitable medium containing the scaffold material, and incubated for a period of time allowing them to proliferate and penetrate the scaffold. In a study by Vunjak-Novakovic et al. it was demonstrated that the requirements for cell proliferation can be easily satisfied by the use of well-mixed spinner flasks during the incubation. In their experiments all cells attached to the scaffolds and no cells were damaged in the process [37]. However, manufacturing polymer scaffolds and manually seeding them with cells has some key limitations. Seeding is not always uniform so the scaffolds need to incubate for up to 72 hours in a bioreactor. Also, cells do not always fully penetrate the entire depth of the scaffold resulting in a non-homogenous distribution of cells. This is not usually an issue in simple, thin constructs but when thick more complex scaffolds are used, such as when attempting to grow a whole organ, this becomes an issue [36].

To overcome the issue of full-depth cell penetration 'cell electrospinning' can be employed. This technique encapsulates living cells into the electrospun fine composite threads. Jayasinghe et al. demonstrated this method, successfully electrospinning living organisms into Poly (dimethylsiloxane) (PDMS) scaffolds [35]. In the study, a coaxial electrospinning setup was used with an inner lumen delivering a suspension of living cells and an outer lumen delivering a PDMS solution. This work suggests the possibility of incorporating living cells into polymer scaffolds with full-depth penetration using electrospinning. This is a significant achievement which has yet to be replicated using any other jet-based techniques [38]. This work from Jayasinghe's group clearly demonstrates the ability to functionalise scaffolds with living cells.

2.2.3.3. *Free surface*

Although needle nanometre diameter fibres of polymer, produced by electrospinning allows excellent control over both fibre diameter and their composition it has an extremely low throughput where basic systems are often limited to flow rates of less than 0.5 mL per hour. A free-surface electrospinning setup such as the El Marco NanoSpider™ Lab 200 system is capable of forming fibres with throughput many thousands of times greater than the conventional needle-based electrospinning setup [39]. In the bowl-based electrospinning approach the spinning solution is simply held in a bath, rather than being delivered through an aperture, with the whole bath then being connected to a high voltage power supply. In the specific case of the NanoSpider™ a rotating metal mandrel is half-submerged in the bath to concentrate the electric field on the thin layer of polymer which coats the mandrel. In this process, many Taylor cones are formed on the surface of the polymer solution, and electrospinning upwards onto a collector above the bath. This increases the throughput of the process many thousands of times above the conventional needle-based system, however much higher voltages, up to 82 kV in the case of the NanoSpider™ are required and solution properties such as viscosity, conductivity and surface tension must be more tightly controlled [40, 41] (**Figure 3**).

2.2.3.4. *Multiple spinnerets*

An electrospinnning setup consisting of multiple spinnerets can allow scale-up from a single needle system. This is a relatively simple setup and can also allow different materials to be mixed during the electrospinnning process. The main disadvantage of the technique is the complicated interactions between the jets. However, it can provide scale-up from a single needle system with similarly tight size distributions. The scale-up is generally linear with the limit being set by the number of needles used.

2.3. The commercial state of electrospinning

The emergence of more scalable electrospinning techniques in recent years has allowed for electrospinning to be used as a commercial fabrication method for non-wovens. For example,

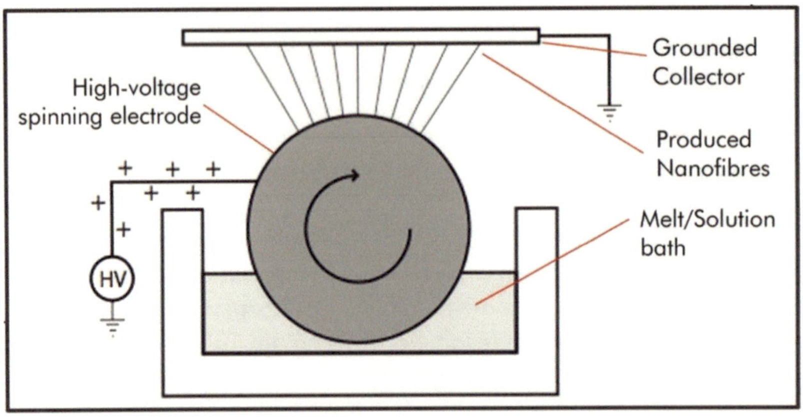

Figure 3. A schematic diagram showing a free-surface electrospinning setup. A polymer solution/melt is held in a bath and a spinning electrode connected to a high voltage power supply is utilised to form multiple jets. Nanofibers are electrospun upwards and collected on a grounded collector plate.

El Marco offer a number of devices, from a lab scale system, NS Lab 200, with a production rate of up to 5.5 g/hour to the NS8S1600U, with a production rate of up to 352 g/hour. These production rates are all significantly higher than for needle electrospinning, and the design of the nanospider devices allows for their incorporation into a production line. This opens up applications with high volume but low value which previously were not feasible using needle electrospinning. These include filtration materials, textiles and dressing materials.

Despite the efforts of researchers to commercialise electrospinning, the upscale is still a significant barrier for many applications. Companies such as Espin technologies (USA) and Fintex Inc. (Korea) are producing nanofibres commercially for specific applications, including air filtrations. Companies such as Nanofibre solutions (USA) and The Electrospinning Company (UK) have been using electrospinning technology to produce nanofibres for niche biomedical applications. There are also companies seeking to take advantage of the properties of nanofibers for a wide range of versatile applications. These companies include Revolution Fibres Ltd. (New Zealand) and Fibre Rio Technology Corp (USA) [42].

There are commercially available nanofibre products in a number of different industries. These include filtration (companies include AMSOIL, Carcor and Donaldson in the air filtration market and Donaldson, DuPont and Finetex in the liquid filter market), acoustic technologies, skin care and biomedical products (companies include Arsenal Medical, The Electrospinning Company and Nanofibre Solutions), Composite materials and sensing and electronics. A more detailed discussion on the companies providing these products and their applications can be found with their commercial applications, challenges and opportunities examined [42].

3. Functional morphology

The morphology of nanofiber mats has many proven benefits for processes including the control of bio interface systems within regenerative medicine. These benefits are innate to

http://dx.doi.org/10.5772/intechopen.73677

nanoscale materials and the high surface to area ratio impacts regenerative medicine in many way including the increased porosity and absorbance of wound dressings of nanofiber mats compared to those formed by micron fibres [43, 44], the biomimicry of the extracellular matrix (ECM) of nanofiber tissue engineering scaffolds [45, 46] and improved control of drug release rates when moving from micro to nanoscale fibres [7, 47]. The following section considers the improved functionality of nanofiber mats compared with materials of more traditional morphology and discusses some recent observations on the impact of application of nanofibers on cellular integration and control.

3.1. Control of mammalian cells

Nanofiber-based scaffolds are of great interest in tissue engineering applications. Their usefulness has been extensively assessed both through *in vitro* and *in vivo* settings. The selection of biocompatible polymers, the modifications to fibre production and processing has been examined in great detail. Of particular interest are the changes to the use of crosslinking agents from cytotoxic chemicals to novel methods which avoid their use. We also assess how structural and orientation changes to the fibres allow the scaffold to fit into a specific niche in the body while the polymer of choice also compliments these changes. Here we will briefly review the recent findings in relation to fibre interactions with mammalian cell systems and the modifications which have been made in order to improve their usefulness in tissue engineering.

One of the largest challenges with electrospun scaffolds is not simply the degree of porosity, which is usually in the region of 91.6% [45], but how accessible the pores are for cell infiltration. Wang et al. propose the use of the Darcy permeability coefficients to determine how cells will respond to surfaces and the architecture of tissue engineering scaffolds through their porosity [48]. They also describe 3 types of pore shown in **Figure 4**; a blind end pore leads to a section of scaffold where material cannot infiltrate further. A closed pore is isolated inside the scaffold and is therefore inaccessible to cells. The desirable architecture is an open pore network where the pores branch between each other and allow cell migration to occur.

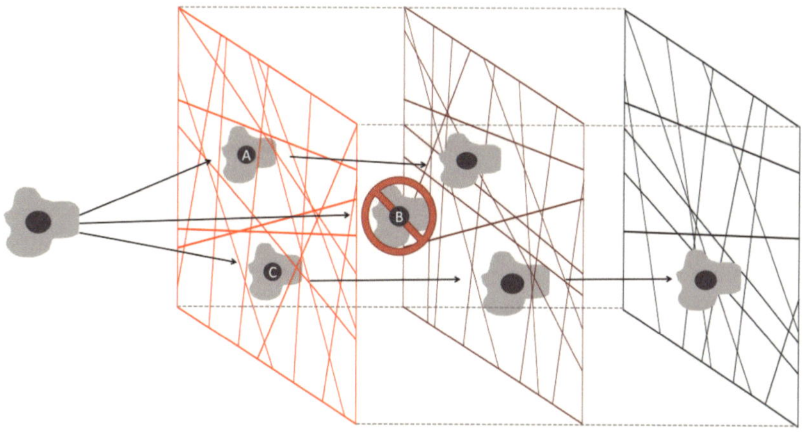

Figure 4. Cell migration through an electrospun scaffold can be difficult due to 3 types of pore: A: Blind end pore where cells cannot completely infiltrate. B: A closed pore inside the scaffold cannot be accessed by cells. C: Open pore network allowing complete cellular infiltration throughout the scaffold.

Li et al. assessed the state of nanofibrous scaffolds for tissue engineering, with great focus on the technology of electrospinning [49]. This review looked into the ideal characteristics for a tissue engineered scaffold, and the questions which should be answered for a particular niche. These begin with architecture; are the fibres of aligned or random orientation? The porosity of a scaffold which must allow for cell migration and nutrient diffusion, and if a scaffold is to recreate blood vessels, are red blood cells contained? This is one of the potential downfalls for an electrospun scaffold, as pore size cannot be easily controlled during production. To avoid this, various methods such as salt leaching have been employed to increase and control pore size [50]. The mechanical properties of a scaffold must also be assessed to match the niche it will be integrated into as these can have drastic effects on cell morphology, cell proliferation and differentiation [51].

Polymer choice must be considered when trying to create a nanofiber scaffold for tissue engineering, especially if the construct is to become a medical device for use in the body. This polymer will need to comply with rigorous testing, and its production as well as the device fabrication will have to follow ISO 13485 standards ("ISO 13485:2016 – Medical devices – Quality management systems – Requirements for regulatory purposes," 2016), followed by European Union Directives 93/42/EEC, 90/385/EEC and 98/79/EEC conformity. As well as considering the polymers safety aspect, it must also be useful to the environment in which it will occupy. Tissue engineering biomaterials can either be synthetic polymers such as poly (lactic-co-glycolic) acid, or naturally occurring polymers; collagen, complex sugars such as hyaluronan or chitosan, or inorganics like hydroxyapatite all fit within this classification [53]. These can be further adapted with surface modifications [54, 55], drug loading [54, 56] and other techniques to increase scaffold efficiency.

Naturally occurring polymers are often favoured, in particular extracellular matrix derived scaffolds such as collagen, which provides support for cell attachment *in vivo* and has been of interest in tissue engineering for over 20 years. The issue with naturally derived polymers tends to be in their processing, many of these cannot be electrospun without the use of harsh solvents such as 1,1,1,3,3,3 Hexafluoro-2-Propanol (HFP) and 2,2,2-Trifluoroethanol (TFE) which have been shown to denature the proteins [57], removing many of the beneficial aspects of their use.

For physiologically soluble polymers it is usually required that the scaffold is crosslinked. Glutaraldehyde has for many years been regarded as the standard method of crosslinking [58], particularly with proteins due to its efficiency and high degree of crosslinking across different polymers [59]. This chemical has many setbacks with regard to tissue engineering scaffolds, foremost the calcification of scaffolds and surrounding tissues which are exposed to residual crosslinking agent which may lead to device failure [60]. In recent years, alternatives to glutaraldehyde have arisen with the desire to increase cytocompatibility *in vivo*, examples of these can be seen in **Table 1**.

Niu et al. compared PCL/Collagen scaffolds which were crosslinked with either glutaraldehyde vapour or genipin and examined their efficacy with cell infiltration, survival and proliferation (Niu et al., [58]). They found that nanofiber-based scaffolds increased cell proliferation while microfibre scaffolds showed better infiltration of cells. They also showed that genipin

Crosslinking agent	Example polymers for use	Type
1-ethyl-3-(3-dimethylaminopropyl)carbodiimide hydrochloride (EDC)	Any protein	Chemical
Genipin	Any protein	Chemical
Rose Bengal and 532 nm excitation	Collagen	Photochemical
Citric Acid	Zein, Collagen	Chemical
Thermal cycling induced crystalisation	PVA	Thermal
Pentaerythritol triacrylate (PETA) and UV irradiation	PEO	Photochemical
Lysyl oxidase	Collagen	Native Enzymatic

Table 1. Examples of crosslinking agents used for physiologically soluble polymers in electrospinning applications to increase cytocompatability in comparison with glutaraldehyde and their corresponding polymers.

showed higher levels of cytocompatibility than scaffolds crosslinked with glutaraldehyde. The increased use of less cytotoxic crosslinking agents shows a shift in the field away from the use of compounds that undo the beneficial effects of electrospun scaffolds.

Finally, we will examine the methods used to modify the architecture of the scaffold to best suit its desired niche. The work in this area mainly focuses on the creation of aligned fibres. Electrospinning typically produces a mat which contains randomly orientated 'non-woven' fibres. This type of mat is useful for creating barrier systems such as skin, however many niches are composed of a highly ordered structure in order to direct cell growth and mechanical strength. A prime example of this requirement is muscle tissue regeneration. Avis et al. produced aligned fibres of PLGA by electrospinning, where no further modification was needed [61]. The C2C12 murine myoblasts which were used in this study showed good adhesion and proliferation to the fibres and the RPM 1500 group showed good alignment along the fibre direction where nonaligned RPM 300 group showed a swirling pattern. This helps to demonstrate the importance the control over the scaffold architecture has on the course of scaffold efficacy. It should be noted however that this study demonstrates again that scaffold infiltration is hindered by poor internal access to pores in electrospun samples.

To conclude, it is important that when planning the use of electrospun fibres in applications utilising mammalian cells, there should be consideration of the type of niche the scaffold will be emulating or integrating with. The control over architecture, porosity and mechanical strength must be considered since each component characteristic can drastically affect the efficacy of the scaffold. We must also consider the post-processing that we carry out to improve the scaffold, and whether the current methods are sufficient to emulate a given niche. The move away from glutaraldehyde seeks to demonstrate that there are many minor changes to our processing which may improve the scaffolds use in tissue engineering.

3.2. Control of microbial cells

The growing use of electrospinning to fabricate nanofibrous structures for use in wound dressings, tissue engineering and filtration processes has increased the need for an understanding of the interactions between bacteria and nanostructures. The adhesion characteristics and

Polymer	Solvent	Fibre diameter	Ref
Drug delivery system			
(a) Poly(ε-caprolactone) (shell) + poly(ethylene glycol) (core)	2-2-2-Trifluoroethanol (b) water	200–350 nm	[68]
(a) Poly(ε-caprolactone) and poly(ethylene glycol) (shell)-dextran (core)	Chloroform and DMF-water	1–5 μm	[69]
Poly(ε-caprolactone) (shell)- poly(ethylene glycol) (core)	Chloroform and DMF-water	500–700 nm	[33]
Poly(ε-caprolactone-co-ethyl ethylene phosphate)	DCM and PBS	~4 μm	[70]
Poly(d-l-lactic-co-glycolic acid)- PEG-b-PLA and PLA	DMF	260–250 nm	[71]
Poly(d-l-lactic-co-glycolic acid)	DCM	1–10 μm	[72]
Poly(d-l-lactic-co-glycolic acid)	THF:DMF	400–600 nm	[73–75]
Poly(l-lactide-co-glycolide) and PEG-PLLA	Chloroform	690–1350 nm	[76]
General tissue engineering			
Poly(ε-caprolactone)	Chloroform and methanol	2–10 nm	[77]
Poly(ε-caprolactone) (core) + zein (shell)	Chloroform and DMF	500–900 nm	[78]
Poly(ε-caprolactone) (core) + collagen (shell)	2–2-2-Trifluoroethanol	500 nm	[79]
Poly(d-l-lactic-co-glycolic acid) and PLGA-b-PEG-NH2	DMF and THF	400–1000 nm	[80]
Poly(d-l-lactide-co-glycolide)	DMF and THF	500–800 nm	[81]
Poly(ethylene glycol-co-lactide)	DMF and acetone	1–4 mm	[82]
Poly(ethylene-co-vinyl alcohol)	2-Propanol and water	0.2–8.0 mm	[83]
Collagen	HFP	180–250 nm	[84]
Gelatin	2-2-2-Trifluoroethanol	0.29–9.10 mm	[85]
Fibrinogen	HFP	120–610 μm	[86]
Poly(glycolic acid) and chitin	HFP	130–380 nm	[87]
Vascular tissue engineering			
Poly(ε-caprolactone)	Chloroform and DMF	0.2–1 nm	[88]
Poly(l-lactide-co-ε-caprolactone)	Acetone	200–800 nm	[89, 90]
Poly(propylene carbonate)	Chloroform	5 μm	[91]
Poly(l-lactic acid) and hydroxyapatite	DCM and 1-4-dioxane	300 nm	[92]
Chitin	HFP	0.16–8.77 nm	[93]

Table 2. A table showing polymers used for electrospinning and their targeted applications. Reproduced with permission from [84].

colonisation of bacteria on these materials is still not completely understood but is essential to aid their future development [62].

Although there has been a vast amount of research on the interactions between nanofibers and eukaryotic cells [63, 64] there has been very little research focussed on the interaction of micro-organisms with nanofibers [45]. An improved understanding of the fundamental processes involved in such interactions is essential to aid the further development of nanofiber mats for application in environments such as wound dressings, filtration and tissue engineering that can all be compromised by microbial colonisation.

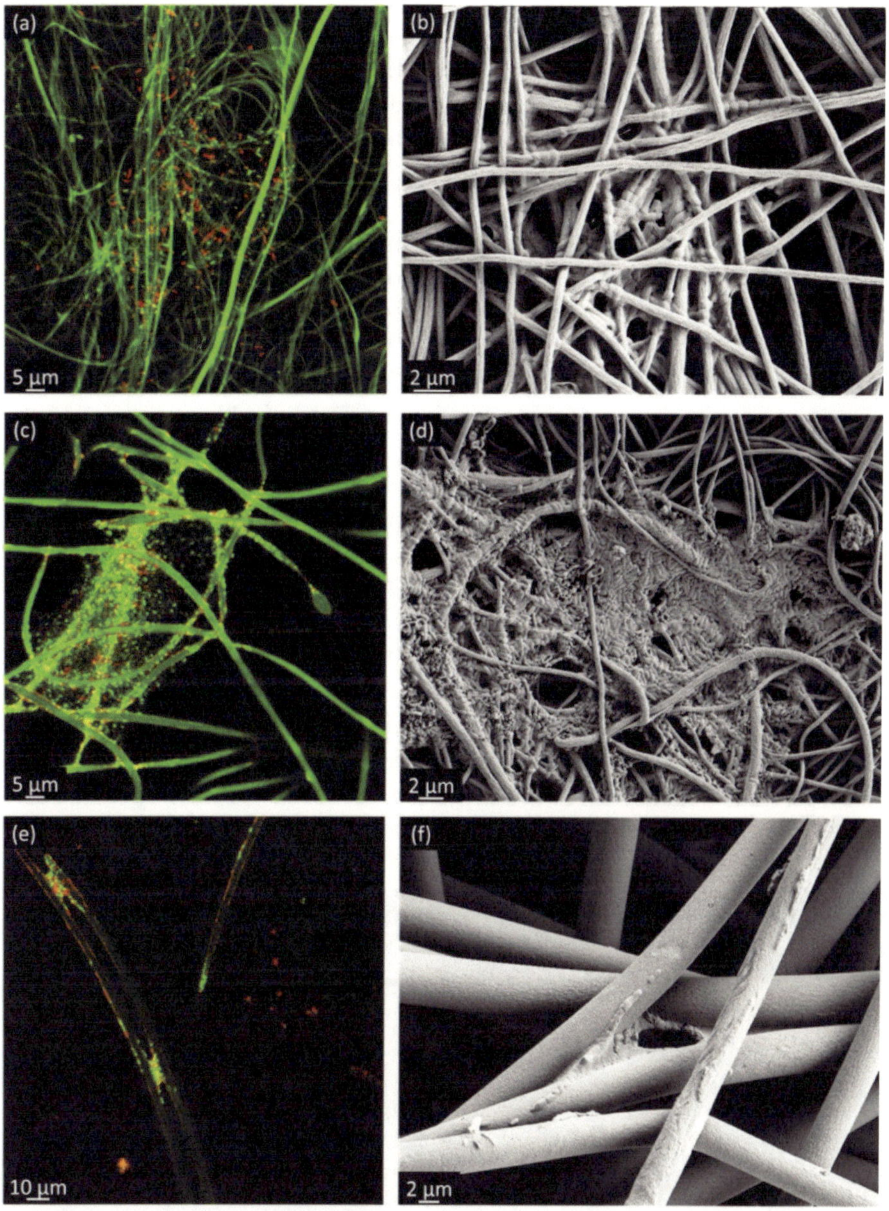

Figure 5. Confocal (a, b and c) and SEM (d, e and f) images of *E. coli* cells interacting with electrospun polystyrene meshes with varying fibre diameters. (a,b) Average diameter = 500 ± 200 nm; (c, d) average diameter = 1000 ± 100 nm; (e, f) average diameter = 3000 ± 1000 nm. Reprinted with permission from [65]. Copyright 2015 American Chemical Society.

Abriago *et al.* investigated the effect of fibre diameter on bacterial attachment, proliferation and growth at electrospun fibre constructs (**Figure 4**) [65]. In their study varying concentrations of polystyrene (PS) in DMF were used to electrospin fibres of different diameter from 300 nm to 3000 nm. Electrospun meshes were then tested against *E. coli, P. aeruginosa* and *S. aureus* both in solution and on agar plates. Their work demonstrated that the fibre diameter influences bacterial proliferation. An average fibre diameter close to that of the bacteria offered the best support for bacterial adhesion and proliferation. Rod shaped cells tended to wrap themselves around fibres with a smaller diameter than their length limiting the ability of the cells to bridge gaps between fibres and form colonies (**Figure 5**). Round cells tended to proliferate through nanofibrous substrates yet when the diameter was larger they were found to have adhered to the surface. Again, these findings were limited by the absence of a smooth control samples. In a further study Abriago et al. investigated the bacterial response to different surface chemistries of electrospun nanofibers [66]. Polystyrene (PS) nanofibers were electrospun and plasma coated with a number of different monomers including allylamine (ppAAm), acrylic acid (ppAAc), 1,7-octadiene (ppOct) and 1,8-cineole (ppCo). The same techniques as the previous paper were used to characterise bacterial interactions with the fibres [65]. The plasma coating did not induce a significant change in fibre morphology. The surface chemistry was found to have a significant effect on bacterial adhesion and proliferation. A ppAAm coating (hydrophilic and rich in amine positively charged groups) resulted in the highest attraction of viable *E. coli* cells forming colonies and clusters across the interstices of the mesh. There was a significantly lower number of *E. coli* cells found on fibres with a hydrophilic, negatively charged ppAAc coating. The cells spread throughout the fibrous network. Fibres with a hydrophobic ppOct coating were found to have a higher proportion of live cells when compared to untreated PS fibres forming clusters at fibre crossovers. The ppCo coating had no inhibitory effect although a high proportion of dead isolated bacterial cells were found to have adhered to the fibres. Cells were wrapped around fibres with no clusters at fibre crossovers or across the interstices of the mesh. The results demonstrate the effect of surface chemistry on the interaction between bacteria and nanofibers offering a further parameter to be altered during electrospinning fibre fabrication to meet a specific antimicrobial attachment application (**Figure 6**).

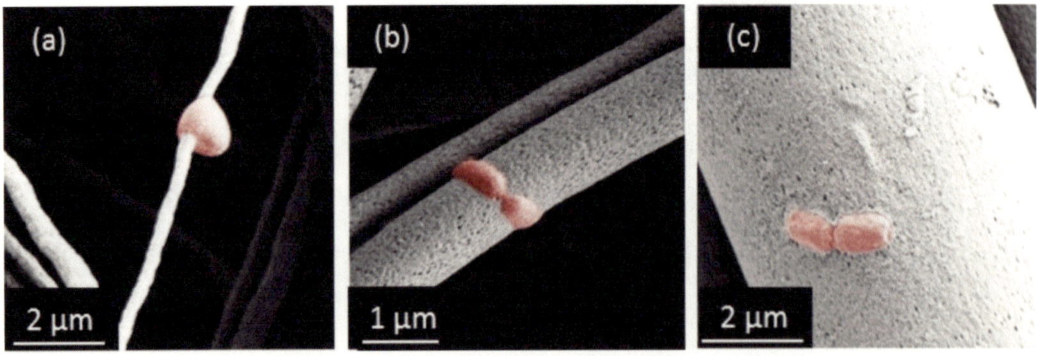

Figure 6. SEM images of *E. coli* cells (red) adhered to polystyrene fibres of average diameter (a) 0.3 μm (b) 1 μm (c) 5 μm. Reprinted with permission from [65]. Copyright 2015 American Chemical Society.

4. Functional materials

There are a wealth of materials that have been electrospun for regenerative medicine and the choice of material is governed by the application. When choosing materials for tissue engineering, biocompatibility and biodegradability are the dominant functionalities that are required, so this has led to the electrospinning of natural materials such as collagen, gelatin and fibrinogen. However, concerns over contamination and interspecies transfer of disease agents has meant that other biodegradable materials have been considered and these include the aliphatic polyesters such as polylactic acid (PLA), polyglycolic acid (PGA), their copolymers (e.g. PLGA) and polycaprolactone (PCL). Electrospinning has an advantage of relatively easy incorporation of biologically active materials into the nanofiber construct. Obviously, this can be immobilisation after fibre fabrication, however there is scope to incorporate the material during electrospinning to create a one-step process, with economic benefit when considering scale-up, and possible mechanisms for controlled release, if required. Incorporation during the fibre fabrication stage may mean a homogeneous distribution of the material throughout the fibre construct as opposed to heterogeneous distribution at interfaces when the fibres are functionalised after fabrication. A caveat for the advantages bestowed by adding functionality through the addition of bioactive materials during the electrospinning process is the potential loss of activity due to location of the additives within the core of the fibre or compromise of the material because of the presence of aggressive solvents used in the electrospinning process. We now discuss the development and application of different materials that confer increased functionality onto the electrospun scaffold.

Material selection is critical in the design of nanofibrous scaffolds for biomedical engineering. The materials suitable for these applications are classified as "biomaterials". A common class of materials used for scaffold fabrication is that of polymers, defined as a large molecular chain composed of multiple repeated subunits. These can be both naturally occurring, "biopolymers", as well as man-made, "synthetic" polymers. The polymer size can be expressed in terms of its molecular weight which is directly related to its average chain length.

The human body's extracellular matrix is composed of natural polymers, primarily polysaccharides and glycosaminoglycan's [46]. Examples of biopolymers commonly used in artificial ECM production are collagen, gelatin, elastin, fibrinogen and chitosan. Examples of commonly used synthetic polymers are Poly (vinyl alcohol) (PVA), Poly (lactic acid) (PLA), Poly (vinylpyrrolidone) (PVP), Poly (lactic-co-glycolic) acid (PLGA) and Poly (ethylene oxide) (PEO). Biopolymers tend to be more biocompatible but have a lower tensile strength compared to synthetic polymers which are generally less biocompatible.

Polymers are generally processed in liquid form as either solutions or melts, with the choice of solvent (or lack thereof) depending on the polymer being used, its solubility and its intended application. The general chemistry rule is 'Like dissolves like', which means that only 'non-polar' solvents will dissolve 'non-polar' solutes and 'polar' solvents dissolve 'polar' solutes. As already mentioned, natural polymers show better biocompatibility than synthetic polymers. These include chitosan, gelatin, collagen, fibrinogen and many others. Synthetic polymers offer the advantage of being able to tailor the properties of the polymer for the desired applications. A number of synthetic polymers are currently used in applications such as wound healing and tissue engineering.

Solubility is another challenge faced by engineers when selecting an appropriate polymer. Water solubility is often desirable in applications such as tissue engineering and drug delivery but for application such as filtration water insolubility and chemical stability are crucial parameters. Biodegradable polymers such as PLGA and PCL are often used for tissue engineering applications although these require a harmful chemical to be used as a solvent, which means that any residual needs to be removed before application to ensure it is not toxic. Another alternative is using a water soluble polymer such as PVA or PEO and using a suitable technique to crosslink it to reduce its water solubility. Poly (vinyl alcohol) (PVA) has been used for a number of applications in wound healing, tissue engineering and filtration. PVA is a biocompatible and biodegradable polymer which has a high hydrophilicity, good chemical resistance and is easily processed using a number of different techniques.

The water solubility of PVA presents a problem when required for most applications. This can be addressed by using a suitable method to crosslink the PVA increasing its crystallinity and thus reducing its solubility. This has been demonstrated using a number of methods such as chemical modification with functional groups including dialdehydes [67] and dicarboxylic acids, physical and chemical treatment with heat, irradiation and acid-catalized dehydration.

4.1. Natural materials

4.1.1. Collagen

Collagens are the main group of structural proteins to be found in the extracellular matrix (ECM), and are the most abundant protein found in animals, providing between 25 and 35% of the whole body protein count. Collagens share the characteristic triple helix structure of Gly-X-Y repeats, where X can be any amino acid, and Y is often proline or hydroxyproline. The individual chains form a left handed helix, and the three chains wind around one another in a right handed super helix [68, 69]. This structure makes the collagen fibres insoluble with high tensile strength [70]. Collagen is a highly versatile biomaterial, which has a wide range of applications, and is particularly suited to a wide range of *in vivo* applications due to its low immunogenicity [71–73]. These *in vivo* applications include wound dressings [74], artificial tissue and organ production [75], cartilage tissue regeneration [76, 77], drug delivery systems [78] and biomedical engineering [79].

In order for collagen to be utilised, it must be obtained in a usable form. It must first undergo extraction from an animal source. The standard procedure for obtaining a usable collagen mixture utilises the technique of solubilisation in acetic acid (AcOH), which can be accompanied by pepsin treatment where necessary to aid in both the speed of extraction and yield of the resulting product [80]. However, the use of pepsin in this process results in collagen which is lacking a significant amount of its telomeric region, and is named 'atelo collagen' as such. These extraction processes have been adapted and modified in many ways, though yield has never been shown to rise above 2% from wet weight or 20% dry weight, and some groups have modified the technique used to determine yield, based on hydroxyproline content to demonstrate small increases in yield more dramatically [81]. The extraction process does not lend itself to be scaled to large production levels, and is usually held back by larger variants of benchtop laboratory scale equipment such as centrifuges and dialysis tubing. The difficulty in extracting collagen from source is why the current market cost per gram of collagen rests around one thousand eight hundred pounds (GBP) per gram of soluble material.

Collagen has been extracted from a range of sources, but is commonly obtained from bovine, porcine and equine sources for *in vivo* use [82]. However, these sources have problems with Bovine Spongiform Encephalopathy (BSE), other Transmittable Spongiform Encephalopathies (TSEs) and potential viral vectors that could be transmissible to humans [83, 84]. More recently, jellyfish have emerged as a source of collagen that is an attractive alternative to existing sources due to a plentiful supply Williams [85] and a safer source through lack of BSE risk and potential viral vectors [86] while exhibiting good performance *in vivo* when compared to bovine sources [87].

Once extracted, collagen can be used to form tissue scaffolds based upon various methods. Often a solution of collagen is lyophilised using freeze drying to form an open architecture based scaffold which is ideal for cell migration. This collagen can also be integrated into a solution of HFP or TFE on its own or as a copolymer and electrospun to form nanofiber scaffolds [88]. In both instances the resulting scaffold must be crosslinked using various chemical [89], enzymatic [90] or photoreactive methods [91] due to the extraction processing of the collagen which renders the material soluble to physiological and acidic conditions.

Electrospinning of soluble collagens provides a suitable way of producing scaffolds which closely mimic the high porosity and surface area often seen in small diameter blood vessels, and has the potential as a good tissue engineering scaffold for the production of three dimensional cell cultures, leading to potential applications such as skin grafts. Research into collagen electrospinning has been under pressure in recent years due to the findings that the primary solvents used to electrospin collagen, namely 1,1,1,3,3,3 Hexafluoro-2-Propanol (HFP) and 2,2,2-Trifluoroethanol (TFE) have been shown to denature collagen when dissolved into either of these solvents [57]. There have been efforts since to try and electrospin collagen using benign solvent mixtures [92] however many of these are unable to replicate the successes in fibre morphology and homogeneity that collagen electrospun out of HFP and TSE produced. As well as the use of solvent being of concern, the quality of the extract of collagen being used also affects how well the solution can be electrospun [93]. This relates to many factors from extraction conditions with critical points which must be addressed. **Figure 7** shows how collagen production can be influenced by a number of extraction conditions.

It is because of these issues that collagen has remained an undesirable polymer for electrospinning and some research has moved onto gelatin electrospinning [94]. This denatured form of collagen is a much more affordable alternative to collagen which is much more easily handled using benign solvents such as acetic acid and phosphate buffered saline (**Figure 8**).

4.2. Nanoparticle-nanofiber composites

The facile nature of the electrospinning process allows for the incorporation of additives such as nanoparticles and antimicrobial agents to form composite fibres. It is a simple process, with dispersions of polymers containing other materials being suitable for electrospinning. We now

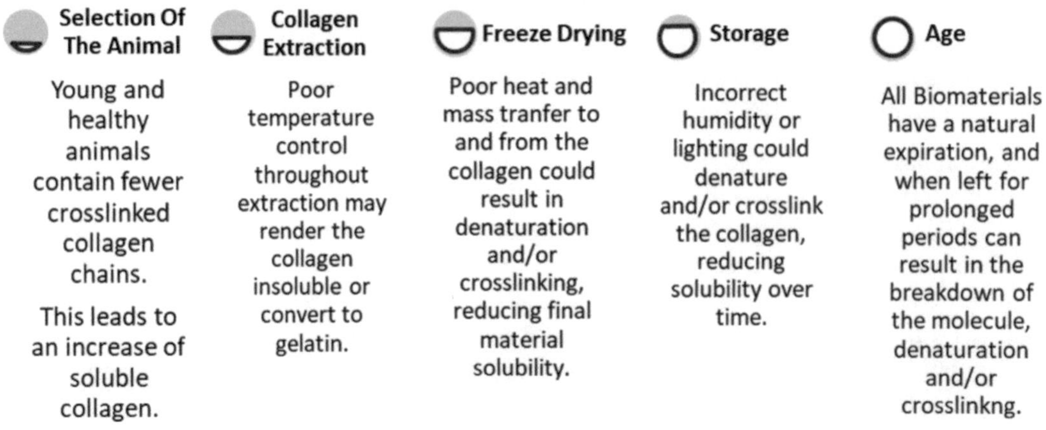

Figure 7. Chart of extraction stages which contain critical points necessary to avoid solubility issues with collagen extracts (Adapted from [93]).

Figure 8. Scanning electron microscopy image of collagen electrospun from HFP. Fibre diameter 107 nm ± 37 nm.

examine the different modes of electrospinning including needle, free-surface and coaxial modes and how these have been used to create composite materials for biomedical engineering.

Electrospun nanoparticle-nanofiber composites have been subject to vast amounts of research in recent times, due to their functionality and unique chemical and physical properties. For example, the incorporation of silver nanoparticles into nanofibers formed from a biocompatible polymer have applications as chronic wound dressings. In another field, tissue engineering, the incorporation of iron oxide nanoparticles into tissue engineering scaffolds has been shown not only to improve the mechanical properties of the polymer nanofibers but also to add further functionality that improves osteogenesis in rabbit models.

There are generally 3 methods commonly used for the fabrication of nanoparticle-nanofiber composites using electrospinning; pre-synthesis, post-synthesis and *in-situ* synthesis. Each of the processes has its advantages, but the most promising technique is *in-situ* synthesis due to its ease of use, simple methodology and scalability.

Nanoparticles are more commonly synthesised before electrospinning or precursors are electrospun and the consequent nanofibers are treated to synthesise the nanoparticles within the nanofibers. More recently, *in-situ* synthesis techniques have emerged which allow for nanoparticles to be synthesised during the electrospinning process or in the solution to be electrospun with no pre-processing of nanoparticles.

The electrospinning process is generally unchanged with all particle synthesis techniques. If particles are pre-synthesised a co-electrospinning technique is generally employed where the nanoparticles are dispersed in a polymer solution before electrospinning. The other technique which can be used is coaxial electrospinning.

Electrospinning nanoparticles which have already been synthesised is the most basic and therefore most commonly used technique for the fabrication of nanoparticle-nanofiber composites. However, this process can often be multi-stage and time consuming requiring particles to be pre-synthesised and subsequently functionalised to reduce the effects of particle agglomeration and allow homogenous distribution throughout the nanofibers. A number of different nanoparticles have been incorporated into nanofibers using this technique including silica [95], titanium oxide (TiO_2) [96], Al_2O_3, Fe_2O_3, SiO_2 and ZnO [97]. Although this technique can be limiting it can also be seen as advantageous for some applications with pre-synthesis of nanoparticles is generally the only suitable method for preparing functionalised nanoparticle-nanofiber composites. Although the author would argue that *in-situ* synthesis techniques show more promise due to their simplicity and scalability, in some fields, including drug delivery pre-synthesis may be the preferred technique. This is due to the requirement of functionalised nanoparticles which is not achievable using emerging techniques.

Post treatment techniques involve the inclusion of a precursor in the electrospun nanofibers which undergo post-electrospinning processing technique to form nanoparticles within the nanofibers. This can also be achieved by immersing nanofibers in a nanoparticle solution, as presented by Razzaz *et al.* with TiO_2 nanoparticles [96], but this requires pre-synthesis of nanoparticles. Other researchers have included a precursor in the electrospun material and synthesised the particles within the fibres. These have included nanoparticles of iron oxide [98] and silver [99]. It could be argued that post-treatment techniques offer no more simplicity that pre-synthesis techniques, although both have their advantages and disadvantages. Generally, pre-synthesis techniques offer no control of the distribution of nanoparticles, although if the nanoparticles are well dispersed the distribution will be reasonably homogenous, throughout the fibres. Post-treatment does not generally allow for this, with encapsulation of nanoparticles difficult. Instead, with post-treatment techniques nanoparticles generally coat the fibres. This tends to result in nanoparticles being released or 'used up' quicker, making nanoparticle-nanocomposites fabricated using a post-treatment technique unsuitable for long term applications but more suitable for applications where a short release cycle is desirable (**Table 3**).

In recent year's research has focussed on the development of *in-situ* synthesis techniques combined with electrospinning with less steps, more simplicity and lower production costs. This is an area that has been investigated in more depth for metal nanoparticles. For example, Jin et al. presented a one-step technique to prepare silver nanoparticles in Poly(vinylpyrrolidone) PVP nanoparticles [100]. In their study silver nitrate ($AgNO_3$) was reduced in a PVP/DMF solution with DMF as the reducing agent. Solutions were then electrospin resulting in PVP nanofibers containing silver nanoparticles. Saquing et al. presented a facile one-step technique to synthesise and incorporate silver nanoparticles into electrospun nanofibers [101]. They chose PEO as the electrospinning polymer which is also used as a reducing again for the metal salt precursor and protects the formed nanoparticles from agglomeration. The fibre quality was improved with the addition of the silver nanoparticles and fibre diameter was reduced due to an increase in the electrical conductivity of the solution. Other researchers have presented similar methods for silver nanoparticles [102] and iron oxide [103] and titanium dioxide [104].

In a recent study we have developed a novel one-stage *in-situ* synthesis technique to fabricate PEO and PVP nanofibers containing magnetite MNPs [105]. We have also demonstrated an ability to scale-up the process from laboratory to industrial scale using a commercially available free-surface electrospinning setup. In our technique, a 2:1 molar ratio of ferric and ferrous chloride is added to a PEO solution in deionised water containing sodium borohydride, used to reduce the ions to nanoparticles. The reaction is allowed to progress before being electrospun (**Figure 4**). Nanofiber mats were crosslinked using UV irradiation, EDX was used to confirm the presence of iron, DLS showed the average nanoparticle diameter to range from 8 nm (PVP) to 26 nm (PEO), XRD confirmed the phase of the nanoparticles to be magnetite and NMR showed a shortening in both T1 and T2 relaxation times confirming the nanoparticles could provide a suitable relaxation channel (**Figure 9**).

Nanoparticle material	Nanofiber material	Application
Zinc oxide	Poly(vinyl alcohol)/sodium alginate	Wound dressing
Silver	Poly(vinyl alcohol)	Wound dressing
Polyethyleneimine-capped silver	Polysulfone	Wound dressing
Silver	Polyurethane	Wound dressing
Titanium dioxide	Poly(vinyl pyrrolidone)	Wound dressing
Titania	Polyurethane	Wound dressing
Titania doped with zinc	Poly(vinyl alcohol)	Wound dressing
Silver	Polylactic-co-glycolic acid	Wound dressing

Table 3. A table showing different nanoparticle-nanocomposite materials with applications in wound dressings.

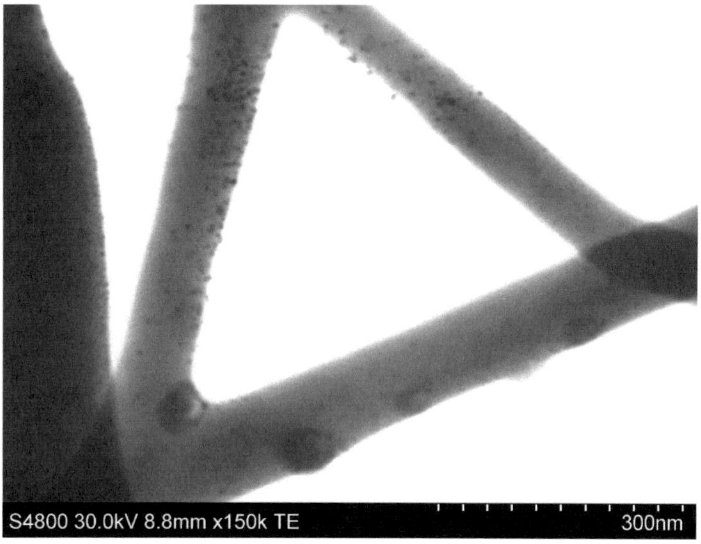

Figure 9. High magnification scanning transmission electron microscopy image of PEO nanofiber containing MNPs.

4.2.1. Applications of nanoparticle-nanofiber composites

Nanoparticle-nanofiber composites have been investigated for a number of different applications in a number of different fields, including antimicrobial applications [106].

Silver nanoparticles are the most commonly used for antimicrobial applications with various silver loaded wound dressings commercially available. Other nanoparticles with antimicrobial properties include copper [107], titanium dioxide (titania) and zinc oxide.

Nanoparticle-nanofiber composites are commonly used as tissue engineering scaffolds to improve mechanical strength, improve hydrophilicity, improve cell migration and proliferation and also for antimicrobial purposes to reduce the competition for colonising stem cells with bacteria. Mehrasa et al. electrospun PLGA/Gelatin scaffolds embedded with mesoporous silica nanoparticles [95]. The incorporation of the nanoparticles was shown to improve both the hydrophilicity and mechanical strength. They also showed improved cell proliferation when compared with pure PLGA scaffolds.

The incorporation of soluble factors and control of surface chemistry of tissue engineering scaffolds to provide biochemical cues have been well documented [108–110]. Magnetic scaffolds have been investigated for the regeneration and repair of tissues in damage and disease [111]. The incorporation of MNPs into scaffolds is also believed to increase *the rate* of both bone cell growth and differentiation. This is due to the tissues ability to recognise the mechano-electrical conversion that can lead to an increased cellular proliferation and expression levels of a number of genes related with bone differentiation [112, 113]. Magnetic scaffolds have also been shown to have applications in tissue engineering [114, 115].

4.3. Wound healing

Electrospinning is becoming a commonly used process in the development of wound dressings with the capability of spinning fibres from a range of both synthetic and natural polymers [7, 44, 116, 117]. They are suitable due to their porosity which allows them to be permeable to water. They are also very absorbent due to their high surface area to volume ratio. The small pore size can be controlled and modified offering semi permeability allowing the wound to stay moist while offering protection from bacteria.

Although they can offer protection from bacteria migrating to the wound they do not protect them from bacteria already present in the wound. Bacteria can still colonise the external surface of the dressing and this can reduce its permeability. For this reason nanofibers are often functionalised with antimicrobial agents [102, 117, 118]. There are a number of dressings on the market which contain antimicrobial agents, with ionic silver being the most commonly used. These offer a number of benefits; not only do they keep the wound aseptic, they can also stop the dressing becoming populated by bacteria.

Silver nanoparticles have been incorporated into nanofibers by a vast amount of researchers. Nguyen *et al.* prepared PVA nanofibers containing silver nanoparticles with applications in wound healing using a combination of microwave irradiation and electrospinning [119]. These were shown to have antimicrobial efficacy against *Staphylococcus aureus* (gram-positive) and *Escherichia coli* (gram-negative), 2 wound relevant organisms. Lakshman et al. also incorporated silver nanoparticles into nanofibers for a wound healing application, using Polyurethane (PU) as the electrospinning polymer due to its current use as an exudate absorptive wound dressing material [120]. The fibres showed a zone of inhibition in a Kirby Bauer disc diffusion assay against *Klebsiella* and was also capable of absorbing 75% of water compared to the control PU sponge.

5. Conclusions

Regenerative medicine is a major focus for 21st century health care and the fabrication of functional nanofibers through electrospinning is an important underpinning technology that is essential if the exciting advances in medicine are to fulfil their potential. As further understanding is achieved for the treatment of disease states by biologically active materials this must be matched by optimisation of electrospinning processes that are able to deliver the functionality to the wound bed or the bodies tissues. Research continues to focus on electrospinning control parameters, which are improving the fundamental understanding of the process with benefit to optimisation strategies and the efficient incorporation of functional materials. This fundamental and optimisation research is also necessary with the recent developments in instrument configurations as the technology of electrospinning emerges from the laboratory bench scale into the realms of economic feasibility and volume manufacture.

Author details

Chris J. Mortimer*, Jonathan P. Widdowson and Chris J. Wright

*Address all correspondence to: c.wright@swansea.ac.uk

Biomaterials, Biofouling and Biofilms Engineering Laboratory (B3EL), Systems and Process Engineering Centre, College of Engineering, Swansea University, Swansea, UK

References

[1] Taylor G. Disintegration of water drops in an electric field. Proceedings of the Royal Society of London A: Mathematical Physics and Engineering Science. 1964;**280**:383-397

[2] Taylor G. Electrically driven jets. Proceedings of the Royal Society of London A: Mathematical Physics and Engineering Science. 1969;**313**:453-475

[3] Yarin AL, Koombhongse S, Reneker DH. Taylor cone and jetting from liquid droplets in electrospinning of nanofibers. Journal of Applied Physics. 2001;**90**:4836-4846

[4] Bhardwaj N, Kundu SC. Electrospinning: A fascinating fiber fabrication technique. Biotechnology Advances. 2010;**28**:325-347. DOI: 10.1016/j.biotechadv.2010.01.004

[5] Reneker DH, Chun I. Nanometre diameter fibres of polymer, produced by electrospinning. Nanotechnology. 1996;**7**:216

[6] T.J. Sill, H.A. von Recum. Electrospinning: Applications in drug delivery and tissue engineering, Biomaterials 2008;**29**:1989-2006. doi: 10.1016/j.biomaterials.2008.01.011

[7] Katti DS, Robinson KW, Ko FK, Laurencin CT. Bioresorbable nanofiber-based systems for wound healing and drug delivery: Optimization of fabrication parameters. Journal of Biomedical Materials Research Part B: Applied Biomaterials. 2004;**70**:286-296

[8] Mit-uppatham C, Nithitanakul M, Supaphol P. Ultrafine electrospun Polyamide-6 Fibers: Effect of solution conditions on morphology and average Fiber diameter. Macromolecular Chemistry and Physics. 2004;**205**:2327-2338

[9] Mo XM, Xu CY, Kotaki ME, Ramamkrishnan S. Ramakrishna, electrospun P (LLA-CL) nanofiber: A biomimetic extracellular matrix for smooth muscle cell and endothelial cell proliferation. Biomaterials. 2004;**25**:1883-1890

[10] Tan SH, Inai R, Kotaki M, Ramakrishna S. Systematic parameter study for ultra-fine fiber fabrication via electrospinning process. Polymer (Guildf). 2005;**46**:6128-6134. DOI: 10.1016/j.polymer.2005.05.068

[11] Andrady AL. Science and Technology of Polymer Nanofibers. Hoboken, NJ, USA: John Wiley & Sons, Inc.; 2008. DOI: 10.1002/9780470229842

[12] Garg K, Bowlin GL. Electrospinning jets and nanofibrous structures. Biomicrofluidics. 2011;**5**:13403. DOI: 10.1063/1.3567097

[13] Son WK, Youk JH, Lee TS, Park WH. The effects of solution properties and polyelectrolyte on electrospinning of ultrafine poly(ethylene oxide) fibers. Polymer (Guildf). 2004;**45**:2959-2966. DOI: 10.1016/j.polymer.2004.03.006

[14] Wendorff JH, Agarwal S, Greiner A. Electrospinning. Weinheim, Germany: Wiley-VCH Verlag GmbH & Co. KGaA; 2012. DOI: 10.1002/9783527647705

[15] Fridrikh SV, Yu JH, Brenner MP, Rutledge GC. Controlling the fiber diameter during electrospinning. Physical Review Letters. 2003;**90**:144502. DOI: 10.1103/PhysRevLett.90.144502

[16] Deitzel JM, Kleinmeyer J, Harris D, Beck Tan NC. The effect of processing variables on the morphology of electrospun nanofibers and textiles. Polymer (Guildf). 2001;**42**:261-272. DOI: 10.1016/S0032-3861(00)00250-0

[17] Buchko CJ, Chen LC, Shen Y, Martin DC. Processing and microstructural characterization of porous biocompatible protein polymer thin films. Polymer (Guildf). 1999;**40**:7397-7407

[18] Baumgarten PK. Electrostatic spinning of acrylic microfibers. Journal of Colloid and Interface Science. 1971;**36**:71-79. DOI: 10.1016/0021-9797(71)90241-4

[19] Jalili R, Hosseini SA, Morshed M. The effects of operating parameters on the morphology of electrospun polyacrilonitrile nanofibres. Iranian Polymer Journal. 2005;**14**:1074-1081

[20] Kidoaki S, Kwon IK, Matsuda T. Structural features and mechanical properties of in situ–bonded meshes of segmented polyurethane electrospun from mixed solvents. Journal of Biomedical Materials Research Part B: Applied Biomaterials. 2006;**76B**:219-229. DOI: 10.1002/jbm.b.30336

[21] Li D, Wang Y, Xia Y. Electrospinning of polymeric and ceramic Nanofibers as Uniaxially aligned arrays. Nano Letters. 2003;**3**:1167-1171. DOI: 10.1021/nl0344256

[22] Deitzel JM, Kleinmeyer JD, Hirvonen JK, Beck Tan NC. Controlled deposition of electrospun poly (ethylene oxide) fibers. Polymer (Guildf). 2001;**42**:8163-8170

[23] Teo WE, Ramakrishna S. A review on electrospinning design and nanofibre assemblies. Nanotechnology. 2006;**17**:R89

[24] Song T, Zhang Y, Zhou T, Lim CT, Ramakrishna S, Liu B. Encapsulation of self-assembled FePt magnetic nanoparticles in PCL nanofibers by coaxial electrospinning. Chemical Physics Letters. 2005;**415**:317-322. DOI: 10.1016/j.cplett.2005.09.035

[25] Teo WE, Kotaki M, Mo XM, Ramakrishna S. Porous tubular structures with controlled fibre orientation using a modified electrospinning method. Nanotechnology. 2005;**16**:918

[26] A. Bornat, Production of Electrostatically Spun Products, EP0009941B2, 1987

[27] Subramanian A, Vu D, Larsen GF, Lin HY. Preparation and evaluation of the electrospun chitosan/PEO fibers for potential applications in cartilage tissue engineering. Journal of Biomaterials Science. Polymer Edition. 2005;**16**:861-873

[28] Teo WE, Ramakrishna S. A review on electrospinning design and nanofibre assemblies. Nanotechnology. 2006;**17**:R89-R106. DOI: 10.1088/0957-4484/17/14/R01

[29] Bellan LM, Craighead HG. Control of an electrospinning jet using electric focusing and jet-steering fields. Journal of Vacuum Science and Technology B. 2006;**24**:3179-3183. DOI: 10.1116/1.2363403

[30] Yang Y, Jia Z, Liu J, Li Q, Hou L, Wang L, Guan Z. Effect of electric field distribution uniformity on electrospinning. Journal of Applied Physics. 2008;**103**:104307. DOI: 10.1063/1.2924439

[31] Sun Z, Zussman E, Yarin AL, Wendorff JH, Greiner A. Compound core-shell polymer nanofibers by co-electrospinning. Advanced Materials. 2003;**15**:1929-1932. DOI: 10.1002/adma.200305136

[32] McCann JT, Marquez M, Xia Y. Melt coaxial electrospinning: A versatile method for the encapsulation of solid materials and fabrication of phase change nanofibers. Nano Letters. 2006;**6**:2868-2872

[33] Jiang H, Hu Y, Li Y, Zhao P, Zhu K, Chen W. A facile technique to prepare biodegradable coaxial electrospun nanofibers for controlled release of bioactive agents. Journal of Controlled Release. 2005;**108**:237-243. DOI: 10.1016/j.jconrel.2005.08.006

[34] Chen H, Wang N, Di J, Zhao Y, Song Y, Jiang L. Nanowire-in-microtube structured Core/Shell Fibers via multifluidic coaxial electrospinning. Langmuir. 2010;**26**:11291-11296. DOI: 10.1021/la100611f

[35] Townsend-Nicholson A, Jayasinghe SN. Cell electrospinning: A unique biotechnique for encapsulating living organisms for generating active biological microthreads/scaffolds. Biomacromolecules. 2006;**7**:3364-3369. DOI: 10.1021/bm060649h

[36] Jayasinghe SN. Cell electrospinning: A novel tool for functionalising fibres, scaffolds and membranes with living cells and other advanced materials for regenerative biology and medicine. The Analyst. 2013;**138**:2215-2223. DOI: 10.1039/c3an36599a

[37] Vunjak-Novakovic G, Obradovic B, Martin I, Bursac PM, Langer R, Freed LE. Dynamic cell seeding of polymer scaffolds for cartilage tissue engineering. Biotechnology Progress. 1998;**14**:193-202. DOI: 10.1021/bp970120j

[38] Jayasinghe SN. Biojets in regenerative biology & medicine. Materials Today. 2011;**14**:202-211. DOI: 10.1016/S1369-7021(11)70115-8

[39] Wang X, Niu H, Lin T, Wang X. Needleless electrospinning of nanofibers with a conical wire coil. Polymer Engineering and Science. 2009;**49**:1582-1586

[40] Thoppey NM, Bochinski JR, Clarke LI, Gorga RE. Edge electrospinning for high throughput production of quality nanofibers. Nanotechnology. 2011;**22**:345301. DOI: 10.1088/0957-4484/22/34/345301

[41] Niu H, Lin T. Fiber generators in needleless electrospinning. Journal of Nanomaterials. 2012;**2012**:12

[42] Kannan B, Cha H, Hosie IC. Electrospinning—Commercial applications, challenges and opportunities. In: Nano-Size Polym. Springer; 2016. pp. 309-342

[43] A.J. Hassiba, M.E. El Zowalaty, G.K. Nasrallah, T.J. Webster, A.S. Luyt, A.M. Abdullah, A.A. Elzatahry. Review of recent research on biomedical applications of electrospun polymer nanofibers for improved wound healing, Nanomedicine; 2016

[44] Abrigo M, McArthur SL, Kingshott P. Electrospun nanofibers as dressings for chronic wound care: Advances, challenges, and future prospects. Macromolecular Bioscience. 2014;**14**:772-792. DOI: 10.1002/mabi.201300561

[45] Li W, Laurencin CT, Caterson EJ, Tuan RS, Ko FK. Electrospun nanofibrous structure: A novel scaffold for tissue engineering. Journal of Biomedical Materials Research. 2002;**60**: 613-621

[46] Sell SA, Wolfe PS, Garg K, McCool JM, Rodriguez IA, Bowlin GL. The use of natural polymers in tissue engineering: A focus on electrospun extracellular matrix analogues. Polymers (Basel). 2010;**2**:522-553 http://www.mdpi.com/2073-4360/2/4/522

[47] Xie J, Wang CH. Electrospun micro- and nanofibers for sustained delivery of paclitaxel to treat C6 glioma in vitro. Pharmaceutical Research. 2006;**23**:1817-1826. DOI: 10.1007/s11095-006-9036-z

[48] Wang Y, Tomlins PE, Coombes AGA, Rides M. On the determination of Darcy permeability coefficients for a microporous tissue scaffold. Tissue Engineering. Part C, Methods. 2010;**16**:281-289. DOI: 10.1089/ten.tec.2009.0116

[49] Li W, Shanti RM, Tuan RS. Electrospinning Technology for nanofibrous scaffolds in tissue engineering. 2006. DOI: 10.1002/9783527610419.ntls0097

[50] Lee YH, Lee JH, An IG, Kim C, Lee DS, Lee YK, Do Nam J. Electrospun dual-porosity structure and biodegradation morphology of Montmorillonite reinforced PLLA nanocomposite scaffolds. Biomaterials. 2005;**26**:3165-3172. DOI: 10.1016/j.biomaterials.2004.08.018

[51] Hubbell JA. Biomaterials in tissue engineering. Biotechnology (NY). 1995;**13**:565-576. DOI: 10.1038/nbt0695-565

[52] International Organization of Standardization, ISO 13485:2016 - Medical devices – Quality management systems – Requirements for regulatory purposes, www.iso.org; 2016. pp. 1-36

[53] Kohane DS, Langer R. Polymeric biomaterials in tissue engineering. Pediatric Research. 2008; **63**:487-491. DOI: 10.1203/01.pdr.0000305937.26105.e7\r00006450-200805000-00006 [pii]

[54] Chung HJ, Park TG. Surface engineered and drug releasing pre-fabricated scaffolds for tissue engineering. Advanced Drug Delivery Reviews. 2007;**59**:249-262. DOI: 10.1016/j.addr.2007.03.015

[55] Ma Z, Mao Z, Gao C. Surface modification and property analysis of biomedical polymers used for tissue engineering. Colloids and Surfaces. B, Biointerfaces. 2007;**60**:137-157. DOI: 10.1016/j.colsurfb.2007.06.019

[56] Zhang G, Suggs LJ. Matrices and scaffolds for drug delivery in vascular tissue engineering. Advanced Drug Delivery Reviews. 2007;**59**:360-373. DOI: 10.1016/j.addr.2007.03.018

[57] Zeugolis DI, Khew ST, Yew ESY, Ekaputra AK, Tong YW, Yung LYL, Hutmacher DW, Sheppard C, Raghunath M. Electro-spinning of pure collagen nano-fibres - just an expensive way to make gelatin? Biomaterials. 2008;**29**:2293-2305. DOI: 10.1016/j.biomaterials.2008.02.009

[58] Niu G, Criswell T, Sapoznik E, Lee S, Soker S. The influence of cross-linking methods on the mechanical and biocompatible properties of vascular scaffold. Journal of Science and Applications: Biomedicine. 2013;**1**:1-7

[59] Barbosa O, Ortiz C, Berenguer-Murcia Á, Torres R, Rodrigues RC, Fernandez-Lafuente R. Glutaraldehyde in bio-catalysts design: A useful crosslinker and a versatile tool in enzyme immobilization. RSC Advances. 2014;**4**:1583-1600. DOI: 10.1039/C3RA45991H

[60] Golomb G, Schoen FJ, Smith MS, Linden J, Dixon M, Levy RJ. The role of glutaraldehyde-induced cross-links in calcification of bovine pericardium used in cardiac valve bioprostheses. The American Journal of Pathology. 1987;**127**:122-130

[61] Aviss KJ, Gough JE, Downes S. Aligned electrospun polymer fibres for skeletal muscle regeneration. European Cells & Materials. 2010;**19**:193-204 doi:vol019a19 [pii]

[62] Mortimer CJ, Burke L, Wright CJ. Microbial interactions with nanostructures and their importance for the development of electrospun nanofibrous materials used in regenerative medicine and filtration. Journal of Microbial and Biochemical Technology. 2016;**8**:195-201. DOI: 10.4172/1948-5948.1000285

[63] Nisbet DRR, Forsythe JSS, Shen W, Finkelstein DII, Horne MKK. Review paper: A review of the cellular response on electrospun nanofibers for tissue engineering. Journal of Biomaterials Applications. 2009;**24**:7-29. DOI: 10.1177/0885328208099086

[64] Mi HY, Salick MR, Jing X, Crone WC, Peng XF, Turng LS. Electrospinning of unidirectionally and orthogonally aligned thermoplastic polyurethane nanofibers: Fiber orientation and cell migration. Journal of Biomedical Materials Research Part A. 2015;**103**:593-603. DOI: 10.1002/jbm.a.35208

[65] Abrigo M, Kingshott P, McArthur SL. Electrospun polystyrene fiber diameter influencing bacterial attachment, proliferation, and growth. ACS Applied Materials & Interfaces. 2015;**7**:7644-7652. DOI: 10.1021/acsami.5b00453

[66] Abrigo M, Kingshott P, McArthur SL. Bacterial response to different surface chemistries fabricated by plasma polymerization on electrospun nanofibers. Biointerphases. 2015;**10**: 04A301. DOI: 10.1116/1.4927218

[67] Gebben B, Van den Berg HWA, Bargeman D, Smolders CA. Intramolecular crosslinking of poly (vinyl alcohol). Polymer (Guildf). 1985;**26**:1737-1740

[68] Gorgieva S, Kokol V. Collagen-vs. Gelatin-based biomaterials and their biocompatibility: Review and perspectives. Biomaterials Applications for Nanomedicine. 2011:17-51. DOI: 10.5772/24118

[69] Myllyharju J, Kivirikko KI. Collagens and collagen-related diseases. Annals of Medicine. 2001;**33**:7-21. DOI: 10.3109/07853890109002055

[70] Gelse K, Pöschl E, Aigner T. Collagens - structure, function, and biosynthesis. Advanced Drug Delivery Reviews. 2003;**55**:1531-1546. DOI: 10.1016/j.addr.2003.08.002

[71] Maeda M, Tani S, Sano A, Fujioka K. Microstructure and release characteristics of the minipellet, a collagen-based drug delivery system for controlled release of protein drugs. Journal of Controlled Release. 1999;**62**:313-324. DOI: 10.1016/S0168-3659(99)00156-X

[72] Sinha VR, Trehan A. Biodegradable microspheres for protein delivery. Journal of Controlled Release. 2003;**90**:261-280. DOI: 10.1016/S0168-3659(03)00194-9

[73] Lynn AK, Yannas IV, Bonfield W. Antigenicity and immunogenicity of collagen. Journal of Biomedical Materials Research Part B: Applied Biomaterials. 2004;**71**:343-354. DOI: 10.1002/jbm.b.30096

[74] Nehrer S, Breinan HA, Ramappa A, Young G, Shortkroff S, Louie LK, Sledge CB, Yannas IV, Spector M. Matrix collagen type and pore size influence behaviour of seeded canine chondrocytes. Biomaterials. 1997;**18**:769-776. DOI: 10.1016/S0142-9612(97)00001-X

[75] Kubota Y, Kleinman HK, Martin GR, Lawley TJ. Role of laminin and basement membrane in the morphological differentiation of human endothelial cells into capillary like structures. The Journal of Cell Biology. 1988;**107**:1589-1598. DOI: 10.1083/jcb.107.4.1589

[76] Hoyer B, Bernhardt A, Lode A, Heinemann S, Sewing J, Klinger M, Notbohm H, Gelinsky M. Jellyfish collagen scaffolds for cartilage tissue engineering. Acta Biomaterialia. 2014;**10**: 883-892. DOI: 10.1016/j.actbio.2013.10.022

[77] Sewing J, Klinger M, Notbohm H. Jellyfish collagen matrices conserve the chondrogenic phenotype in two- and three-dimensional collagen matrices. Journal of Tissue Engineering and Regenerative Medicine. 2015. DOI: 10.1002/term.1993

[78] Wallace DG, Rosenblatt J. Collagen gel systems for sustained delivery and tissue engineering. Advanced Drug Delivery Reviews. 2003;**55**:1631-1649. DOI: 10.1016/j.addr.2003.08.004

[79] Derkus B, Arslan YE, Bayrac AT, Kantarcioglu I, Emregul KC, Emregul E. Development of a novel aptasensor using jellyfish collagen as matrix and thrombin detection in blood samples obtained from patients with various neurodisease. Sensors and Actuators B: Chemical. 2016;**228**:725-736. DOI: 10.1016/j.snb.2016.01.095

[80] Addad S, Exposito JY, Faye C, Ricard-Blum S, Lethias C. Isolation, characterization and biological evaluation of jellyfish collagen for use in biomedical applications. Marine Drugs. 2011;**9**:967-983. DOI: 10.3390/md9060967

[81] Nalinanon S, Benjakul S, Visessanguan W, Kishimura H. Use of pepsin for collagen extraction from the skin of bigeye snapper (*Priacanthus tayenus*). Food Chemistry. 2007; **104**:593-601. DOI: 10.1016/j.foodchem.2006.12.035

[82] Silvipriya KS, Krishna Kumar K, Bhat AR, Dinesh Kumar B, John A, Lakshmanan P. Collagen: Animal sources and biomedical application. Journal of Applied Pharmaceutical Science. 2015;**5**:123-127. DOI: 10.7324/JAPS.2015.50322

[83] Asher DM. The transmissible spongiform encephalopathy agents: Concerns and responses of United States regulatory agencies in maintaining the safety of biologics. Developments in Biological Standardization. 1999;**100**:103-118

[84] Lupi O. Prions in dermatology. Journal of the American Academy of Dermatology. 2002; **46**:790-793. DOI: 10.1067/mjd.2002.120624

[85] Williams J. Are jellyfish taking over the world? Journal of Aquaculture & Marine Biology. 2015;**2**:1-13. DOI: 10.15406/JAMB.2015.2.00026

[86] Song E, Yeon Kim S, Chun T, Byun HJ, Lee YM. Collagen scaffolds derived from a marine source and their biocompatibility. Biomaterials. 2006;**27**:2951-2961. DOI: 10.1016/j.biomaterials.2006.01.015

[87] Widdowson JP, Picton AJ, Vince V, Wright CJ, Mearns-Spragg A. *In vivo* comparison of jellyfish and bovine collagen sponges as prototype medical devices. Journal of Biomedical Materials Research Part B: Applied Biomaterials. 2017:1-10. DOI: 10.1002/jbm.b.33959

[88] Chen Z, Mo X, Qing F. Electrospinning of collagen-chitosan complex. Materials Letters. 2007;**61**:3490-3494. DOI: 10.1016/j.matlet.2006.11.104

[89] Barnes CP, Pemble CW, Brand DD, Simpson DG, Bowlin GL. Cross-linking electrospun type II collagen tissue engineering scaffolds with carbodiimide in ethanol. Tissue Engineering. 2007;**13**:1593-1605. DOI: 10.1089/ten.2006.0292

[90] Torres-Giner S, Gimeno-Alcañiz JV, Ocio MJ, Lagaron JM. Comparative performance of electrospun collagen nanofibers cross-linked by means of different methods. ACS Applied Materials & Interfaces. 2009;**1**:218-223. DOI: 10.1021/am800063x

[91] Cherfan D, Verter EE, Melki S, Gisel TE, Doyle FJ, Scarcelli G, Yun SH, Redmond RW, Kochevar IE. Collagen cross-linking using rose bengal and green light to increase corneal stiffness. Investigative Ophthalmology and Visual Science. 2013;**54**:3426-3433. DOI: 10.1167/iovs.12-11509

[92] Dong B, Arnoult O, Smith ME, Wnek GE. Electrospinning of collagen nanofiber scaffolds from benign solvents. Macromolecular Rapid Communications. 2009;**30**:539-542. DOI: 10.1002/marc.200800634

[93] Zeugolis DI, Li B, Lareu RR, Chan CK, Raghunath M. Collagen solubility testing, a quality assurance step for reproducible electro-spun nano-fibre fabrication. A technical

note. Journal of Biomaterials Science. Polymer Edition. 2008;**19**:1307-1317. DOI: 10.1163/156856208786052344

[94] Sajkiewicz P, Kołbuk D. Electrospinning of gelatin for tissue engineering - molecular conformation as one of the overlooked problems. Journal of Biomaterials Science. Polymer Edition. 2014;**25**:2009-2022. DOI: 10.1080/09205063.2014.975392

[95] Mehrasa M, Asadollahi MA, Ghaedi K, Salehi H, Arpanaei A. Electrospun aligned PLGA and PLGA/gelatin nanofibers embedded with silica nanoparticles for tissue engineering. International Journal of Biological Macromolecules. 2015;**79**:687-695. DOI: 10.1016/j.ijbiomac.2015.05.050

[96] Razzaz A, Ghorban S, Hosayni L, Irani M, Aliabadi M. Chitosan nanofibers functionalized by TiO_2 nanoparticles for the removal of heavy metal ions. Journal of the Taiwan Institute of Chemical Engineers. 2016;**58**:333-343. DOI: 10.1016/j.jtice.2015.06.003

[97] Babapoor A, Karimi G, Khorram M. Fabrication and characterization of nanofiber-nanoparticle-composites with phase change materials by electrospinning. Applied Thermal Engineering. 2016;**99**:1225-1235. DOI: 10.1016/j.applthermaleng.2016.02.026

[98] Xiao S, Shen M, Guo R, Wang S, Shi X. Immobilization of Zerovalent iron nanoparticles into electrospun polymer nanofibers: Synthesis, characterization, and potential environmental applications. Journal of Physical Chemistry C. 2009;**113**:18062-18068. DOI: 10.1021/jp905542g

[99] Wang Y, Yang Q, Shan G, Wang C, Du J, Wang S, Li Y, Chen X, Jing X, Wei Y. Preparation of silver nanoparticles dispersed in polyacrylonitrile nanofiber film spun by electrospinning. Materials Letters. 2005;**59**:3046-3049. DOI: 10.1016/j.matlet.2005.05.016

[100] Jin W-J, Lee HK, Jeong EH, Park WH, Youk JH. Preparation of polymer nanofibers containing silver nanoparticles by using poly(N-vinylpyrrolidone). Macromolecular Rapid Communications. 2005;**26**:1903-1907. DOI: 10.1002/marc.200500569

[101] Saquing CD, Manasco JL, Khan SA. Electrospun nanoparticle-nanofiber composites via a one-step synthesis. Small. 2009;**5**:944-951. DOI: 10.1002/smll.200801273

[102] Rujitanaroj P, Pimpha N, Supaphol P. Wound-dressing materials with antibacterial activity from electrospun gelatin fiber mats containing silver nanoparticles. Polymer (Guildf). 2008;**49**:4723-4732

[103] Faridi-Majidi R, Sharifi-Sanjani N. In situ synthesis of iron oxide nanoparticles on poly (ethylene oxide) nanofibers through an electrospinning process. Journal of Applied Polymer Science. 2007;**105**:1351-1355. DOI: 10.1002/app.26230

[104] Yan L, Si S, Chen Y, Yuan T, Fan H, Yao Y, Zhang Q. Electrospun in-situ hybrid polyurethane/nano-TiO2 as wound dressings. Fibers and Polymers. 2011;**12**:207-213. DOI: 10.1007/s12221-011-0207-0

[105] Burke L, Mortimer CJ, Curtis DJ, Lewis AR, Williams R, Hawkins K, Maffeis TGG, Wright CJ. In-situ synthesis of magnetic iron-oxide nanoparticle-nanofibre composites

using electrospinning. Materials Science and Engineering: C. 2017;**70**:512-519. DOI: 10.1016/j.msec.2016.09.014

[106] Yah CS, Simate GS. Nanoparticles as potential new generation broad spectrum antimicrobial agents. DARU. 2015;**23**:43. DOI: 10.1186/s40199-015-0125-6

[107] Zhang C, Li C, Bai J, Wang J, Li H. Synthesis, characterization, and antibacterial activity of cu NPs embedded electrospun composite nanofibers. Colloid & Polymer Science. 2015;**293**:2525-2530. DOI: 10.1007/s00396-015-3640-6

[108] Rezwan K, Chen QZ, Blaker JJ, Boccaccini AR. Biodegradable and bioactive porous polymer/inorganic composite scaffolds for bone tissue engineering. Biomaterials. 2006; **27**:3413-3431. DOI: 10.1016/j.biomaterials.2006.01.039

[109] Whitaker MJ, Quirk RA, Howdle SM, Shakesheff KM. Growth factor release from tissue engineering scaffolds. The Journal of Pharmacy and Pharmacology. 2001;**53**:1427-1437. DOI: 10.1211/0022357011777963

[110] Lutolf MP, Hubbell JA. Synthetic biomaterials as instructive extracellular microenvironments for morphogenesis in tissue engineering. Nature Biotechnology. 2005;**23**:47-55. DOI: 10.1038/nbt1055

[111] Singh RK, Patel KD, Lee JH, Lee EJ, Kim JH, Kim TH, Kim HW. Potential of magnetic nanofiber scaffolds with mechanical and biological properties applicable for bone regeneration. PLoS One. 2014;**9**. DOI: 10.1371/journal.pone.0091584

[112] Lang SB. Pyroelectric effect in bone and tendon. Nature. 1966;**212**:704-705. DOI: 10.1038/212704a0

[113] Nomura S, Takano-Yamamoto T. Molecular events caused by mechanical stress in bone. Matrix Biology. 2000;**19**:91-96. DOI: 10.1016/S0945-053X(00)00050-0

[114] Bock N, Riminucci A, Dionigi C, Russo A, Tampieri A, Landi E, Goranov VA, Marcacci M, Dediu V. A novel route in bone tissue engineering: Magnetic biomimetic scaffolds. Acta Biomaterialia. 2010;**6**:786-796. DOI: 10.1016/j.actbio.2009.09.017

[115] Meng J, Xiao B, Zhang Y, Liu J, Xue H, Lei J, Kong H, Huang Y, Jin Z, Gu N, Xu H. Superparamagnetic responsive nanofibrous scaffolds under static magnetic field enhance osteogenesis for bone repair in vivo. Scientific Reports. 2013;**3**. DOI: 10.1038/srep02655 http://www.nature.com/srep/2013/130913/srep02655/abs/srep02655.html#supplementary-information

[116] Kang YO, Yoon I-S, Lee SY, Kim D-D, Lee SJ, Park WH, Hudson SM. Chitosan-coated poly(vinyl alcohol) nanofibers for wound dressings. Journal of Biomedical Materials Research Part B: Applied Biomaterials. 2010;**92B**:568-576. DOI: 10.1002/jbm.b.31554

[117] Zahedi P, Rezaeian I, Ranaei-Siadat S, Jafari S, Supaphol P. A review on wound dressings with an emphasis on electrospun nanofibrous polymeric bandages. Polymers for Advanced Technologies. 2010;**21**:77-95

[118] Lakshman LR, Shalumon KT, Jayakumar R, Nair SV. Preparation of silver nanoparticles incorporated electrospun polyurethane nano-fibrous mat for wound dressing. Journal of Macromolecular Science, Part A Pure and Applied Chemistry. 2010;**47**:1012-1018

[119] Nguyen TH, Kim YH, Song HY, Lee BT. Nano Ag loaded PVA nano-fibrous mats for skin applications. Journal of Biomedical Materials Research Part B: Applied Biomaterials. 2011;**96**(B):225-233. DOI: 10.1002/jbm.b.31756

[120] Lakshman L, Shalumon KT, Nair S, Jayakumar R, Nair SV. Preparation of silver nanoparticles incorporated electrospun polyurethane Nano-fibrous mat for wound dressing. Journal of Macromolecular Science, Part A. 2010;**47**:1012-1018. DOI: 10.1080/10601325.2010.508001

Chapter 7

Effect of Barium Titanate Reinforcement on Tensile Strength and Dielectric Response of Electrospun Polyvinylidene Fluoride Fibers

Avinash Baji and Yiu-Wing Mai

Additional information is available at the end of the chapter

http://dx.doi.org/10.5772/intechopen.74662

Abstract

In this study, we used electrospinning to obtain polyvinylidene fluoride (PVDF) fibers reinforced with barium titanate ($BaTiO_3$) and investigated the influence of $BaTiO_3$ concentration on the tensile strength and dielectric behavior of PVDF fibers. X-ray diffraction (XRD) study and infrared spectroscopy revealed that PVDF fibers filled with $BaTiO_3$ possessed higher fraction of ferroelectric β-crystals compared to neat PVDF fibers. Further, incorporation of 40 wt% $BaTiO_3$ within the fibers increased their stiffness and strength by 95 and 38%, respectively. These improvements in tensile properties of $BaTiO_3$ filled PVDF fibers arose from the reinforcement effect of $BaTiO_3$. Also, the dielectric response of the $BaTiO_3$/PVDF fibers was characterized. The effective dielectric constants of PVDF fibers reinforced with $BaTiO_3$ were found to increase consistently with $BaTiO_3$ content at all frequencies. The dielectric loss of the fibers did not show any significant change for all concentrations of $BaTiO_3$ within the fibers.

Keywords: electrospinning, nanofibers, polyvinylidene difluoride (PVDF), barium titanate ($BaTiO_3$), dielectric properties

1. Introduction

There has been substantial recent interest in the development of nanostructured piezo-sensitive composites as they can potentially display combination of desirable physical properties which cannot be obtained in single phase materials [1–3]. Hence, ceramics such as zinc oxide, bismuth ferrite, lead zirconate titanate, barium titanate ($BaTiO_3$) etc. are widely utilized in actuators, sensors and energy storage devices due to their good piezoelectric and ferroelectric

IntechOpen

© 2018 The Author(s). Licensee IntechOpen. This chapter is distributed under the terms of the Creative Commons Attribution License (http://creativecommons.org/licenses/by/3.0), which permits unrestricted use, distribution, and reproduction in any medium, provided the original work is properly cited.

properties [4–6]. However, their inherit brittleness has limited their use for most engineering applications. By contrast, compliant polymeric materials such as polyvinylidene fluoride (PVDF) and its copolymers are easy to process but generally have low piezoelectric coefficient values when compared to ferroelectric ceramics [7, 8]. Thus, the incorporation of a ceramic phase into an electroactive polymeric matrix could yield composites with improved mechanical integrity and piezoelectric characteristics [1, 2].

Traditionally, such composites are fabricated by dispersing micron or submicron sized ferroelectric ceramics into a dielectric polymer matrix. However, this approach yields composite with film thickness greater than the size of the ceramic particles. Consequently, the composites are invariably found to possess low capacitance densities. Some recent studies demonstrate that reducing the size of the filler particles to nanometer length scales leads to an improvement in the dielectric permittivity of the composite [3, 9]. This is attributed to an increase in the interfacial area which promotes the exchange couple effect, and improves the polarization levels of the composite and its dielectric response [3, 9]. Further, these studies demonstrate that electromechanical coupling and permittivity of the composite can increase by 60 times when large aspect ratio fillers are used and when they are aligned along the poling direction within the matrix phase [10, 11]. Hence, composites reinforced with nanorods or nanowhiskers can display better dielectric permittivity and mechanical strength compared to composites obtained by simply dispersing ceramic powders within a polymer matrix [12–15].

It is well-known that these piezo-sensitive composites are often subjected to external stimuli such as mechanical stresses, electrical field or coupled electromechanical loads [16]. The induced electric field activates the mechanical stresses and the induced mechanical load generates the electrical field within the piezoelectric material [16]. Since the electric field and mechanical stresses interfere with each other, it is therefore of the utmost importance that the composites have good combined mechanical and piezoelectric responses. In our recent study, we fabricated $BaTiO_3$ reinforced PVDF fibers using electrospinning and characterized their piezoresponse [2]. We showed that these PVDF/ $BaTiO_3$ fibers have tremendous potential for future nanoscale electronic devices. Similarly, other researchers working on PVDF/$BaTiO_3$ composites focused their efforts to characterize the dielectric, piezoelectric and ferroelectric behaviors of the composites [7, 14, 15, 17].

In this work, the tensile deformation behavior of PVDF/$BaTiO_3$ fibers obtained using electrospinning is characterized. Their deformation mechanisms are discussed based on their microstructural evolution, such as crystalline structure development induced by the presence of $BaTiO_3$ within the fibers. Finally, results on dielectric permittivity of these nanofibers are presented.

2. Experimental work

2.1. $BaTiO_3$ fibers

We used sol-gel based electrospinning to prepare barium titanate fibers. Briefly, 5.1 g of barium acetate was dissolved in 12 ml of glacial acetic acid; and to this solution, 5.9 ml of titanium isopropoxide was added to obtain $BaTiO_3$ precursor solution [18]. The solution for electrospinning was prepared by dissolving 2.5 g poly(vinyl pyrrolidone) (PVP, MW = 360,000) in 11 ml of solvent solution which was prepared by mixing equal parts of dimethyl formamide (DMF) and ethanol together. Then, $BaTiO_3$ precursor solution was added to the PVP solution.

Electrospinning was conducted using the $BaTiO_3$/PVP solution at 20 kV with 0.07 mm/min as the solution feed rate. The spacing between needle and grounded metal collector was 15 cm. A vacuum oven at 100°C was used for 1 h to dry the fibers. After which the fibers were put in a furnace and annealed at 750°C for 1 h. The resultant fibers, referred to as Sample 1, were $BaTiO_3$ fibers.

2.2. $BaTiO_3$ reinforced PVDF fibers

PVDF fibers filled with 0, 10, 20 and 40 wt% of $BaTiO_3$ were obtained as described in the following steps. In the first step, Sample 1 fibers of known content were dispersed into DMF solution. The solution was sonicated for 0.5 h and then stirred for 1 h to obtain a slurry solution. 18 wt% of PVDF powder was added to the slurry solution for electrospinning.

Electrospinning was conducted at 18 kV on this solution with a feed rate fixed at 0.12 mm/min. PVDF fibers filled with 0, 10, 20 and 40 wt% of $BaTiO_3$ were obtained. This set of $BaTiO_3$-PVDF fiber samples are referred to as Sample 2.

2.3. Microstructure characterization

A scanning electron microscope (FESEM, Zeiss ULTRA plus) was used to observe the microstructure of Sample 1 and Sample 2 fibers. The surface of the samples were gold-coated with a sputter coater before they were examined using SEM. An accelerating voltage of 2–3 kV was used for imaging the samples. Transmission electron microscopy (TEM, Philips CM120 Biofilter) was also used to image Sample 1 and Sample 2 fibers. Sample 2 fibers were electrospun directly on a 400-mesh copper grid and then examined using the TEM.

2.4. X-ray analysis and infrared spectroscopy

The diffraction behavior of Sample 1 and Sample 2 fibers were studied using an X-ray diffractometer (XRD Shimadzu S6000) with Cu $K\alpha$ radiation ($\lambda = 1.54$ Å). The 2θ scan was varied between 15 and 70° and the scan speed was set at 1°/min with 0.02° step size. Bruker Fourier transform infrared spectroscopy system (FTIR, IFS 66v) was used to collect the spectra of Sample 2 fibers. The fibers were scanned from 5000 to 400 cm^{-1} in attenuated total reflectance mode (ATR).

2.5. Mechanical properties

The mechanical integrity of the fibers was analyzed using tensile tests on the aligned fiber samples conducted on an Instron 5567 (2.5 N load cell) testing machine with a cross-head speed of 5 mm/min. The loading direction was parallel to the fiber axis.

Dynamic mechanical measurements on Sample 2 fibers were obtained by a dynamic mechanical analyzer (DMA, TA Instruments). An oscillation amplitude of 10 μm, 3°C/min heating ramp rate and 1 Hz frequency were used.

2.6. Dielectric properties

The frequency-dependent capacitance and loss tangent of Sample 2 fibers as a function of $BaTiO_3$ content were measured using a frequency-response dielectric analyzer (Novocontrol alpha analyzer) with scanning frequencies ranging from 103 to 107 Hz.

3. Results and discussion

Uniform distribution of the ferroelectric ceramic phase within the piezoelectric PVDF matrix is an important prerequisite to obtain electroactive composites with improved physical and mechanical properties, e.g., ferroelectricity, tensile strength, and stiffness [1, 11, 14]. Here, we use electrospinning to obtain PVDF fibers with uniformly dispersed and distributed $BaTiO_3$. The effects of $BaTiO_3$ loading on the structure development and tensile strength of the fibers are evaluated.

The microstructure and size of Samples 1 and 2 are examined using SEM and TEM. **Figure 1A** and **B** show typical SEM and TEM images of Sample 1 $BaTiO_3$ fiber indicating an average fiber diameter of ~170 ± 50 nm. SEM image (**Figure 1A**) reveals that the fibers are composed of fine-grained structures which are assembled and organized to obtain a fibrous geometry. These grains are dense and closely packed as evident in the TEM image (**Figure 1B**). **Figure 2A** and **B** show representative SEM and TEM microstructures of PVDF fibers filled with 20 wt% $BaTiO_3$. An average fiber diameter of Sample 2 (**Figure 2A**) is ~210 ± 40 nm. SEM image shows that the surface morphology of $BaTiO_3$/PVDF fibers appears rough.

Arrows in **Figure 2A** clearly point to slight bulges in the fiber which are owing to the inclusions of $BaTiO_3$ within the PVDF fibers. The inset in **Figure 2A** shows an array of uniaxial aligned and tightly packed fibers which can be easily collected to form a test coupon for tensile experiments. The dispersion quality of $BaTiO_3$ in PVDF can be conveniently examined from the TEM image (**Figure 2B**). Consistent with the SEM images, bulges in the fiber are also noticed (see **Figure 2B**). A magnified TEM image taken from another area of the fiber is given in the inset of **Figure 2B**, which confirms the alignment of $BaTiO_3$ along the fiber axis.

The crystalline phase change in PVDF due to the presence of $BaTiO_3$ is determined using XRD and FTIR. **Figure 3A** shows the XRD patterns of PVDF fibers filled with 0, 10, 20 and 40 wt% $BaTiO_3$. For reference, the XRD pattern of neat $BaTiO_3$ (Sample 1), that has a cubic-tetragonal structure which is responsible for its ferroelectric and dielectric properties [19, 20], is shown in **Figure 3B** revealing the strong peaks. The XRD peaks of Sample 1 fibers displayed in **Figure 3A** match closely the cubic-tetragonal structure of $BaTiO_3$ (JCPDS 31-0174 & JCPDS 05-0626) [2, 19–21]. Sample with 0 wt% $BaTiO_3$ shows 5 peaks corresponding to PVDF crystal structures at 18.3, 20.2, 35.5, 41.1 and 56.1° [7, 22–24]. The peaks at 18.3° and 35.6° (**Figure 3A**) are attributed to the α-crystalline structure of PVDF. The peaks at 20.2, 41.1 and 56.1° are due to the ferroelectric β-crystalline phase of PVDF [2, 7]. Evidently, the main peak for neat

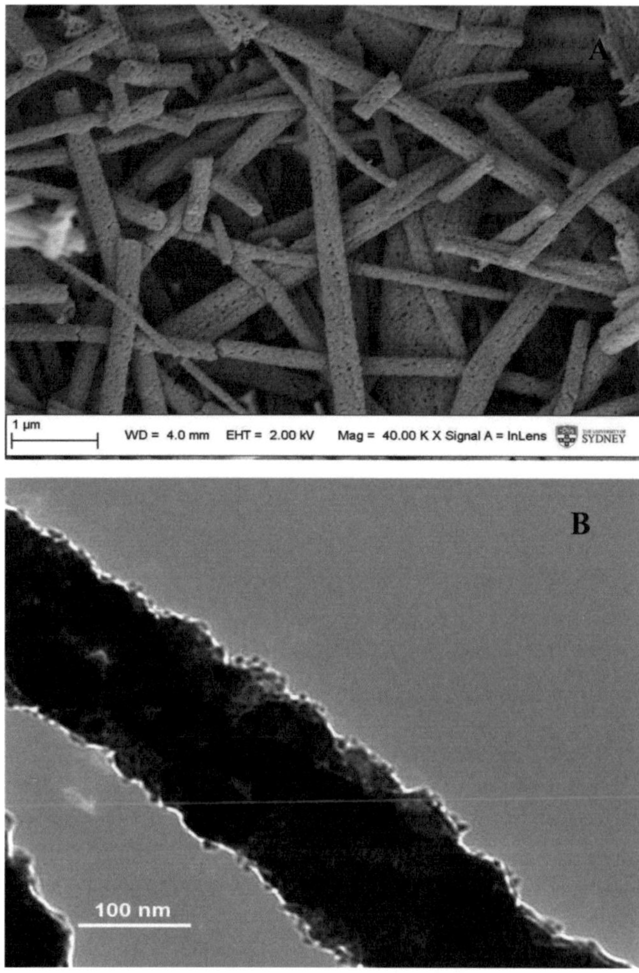

Figure 1. (A) SEM and (B) TEM images of $BaTiO_3$ fiber (Sample 1). The microstructure shows that $BaTiO_3$ grains are self-assembled and self-organized to yield a fibrous geometry.

PVDF is recorded at 20.2°, which is indexed to the 200/110 reflections of the β-crystalline structure. Nonetheless, XRD peaks corresponding to both PVDF and $BaTiO_3$ crystalline phases are clearly shown for the fibers reinforced with 10, 20 and 40 wt% $BaTiO_3$. The intensity of the peak at 20.2° is moderately increased with $BaTiO_3$ content. Furthermore, the intensities of the peaks related to $BaTiO_3$ are seen to increase and become more prominent with increasing filler content in the composite fibers.

The crystalline structure confirmation for the fibers reinforced with $BaTiO_3$ was done aided by the FTIR spectra recorded for the fibers. **Figure 4A** shows the IR spectra of Sample 2 fibers. All the samples show peaks at 766, 840, 1280, 1400 and 1432 cm^{-1}. The bands corresponding to 766, 1400 and 1432 cm^{-1} in the spectra are indexed to the α-crystals of the PVDF matrix, while the bands at 840 and 1280 cm^{-1} in the spectra are indexed to the β-crystals of PVDF [2, 7, 23, 24]. These results indicate that the fiber samples irrespective of the $BaTiO_3$ content contain both α and β

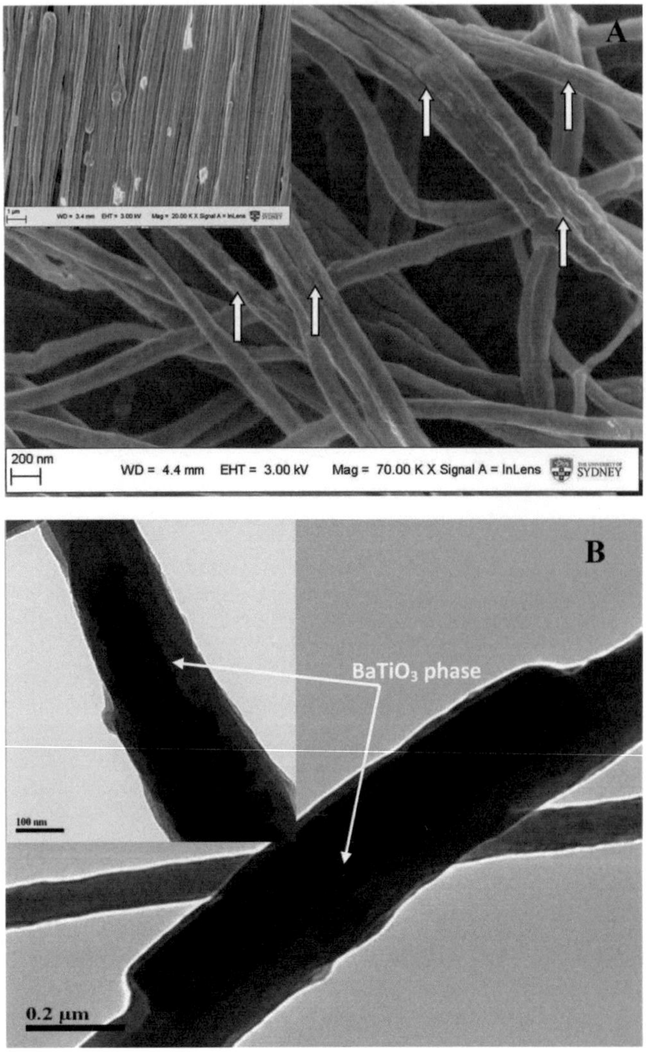

Figure 2. (A) SEM image of PVDF reinforced $BaTiO_3$ fibers (Sample 2). The inset in the figure shows the microstructure of uniaxially aligned fiber arrays. Aligned PVDF fibers as a function of $BaTiO_3$ content were collected for characterizing their tensile properties. (B) TEM images of typical PVDF fibers reinforced with 20 wt% $BaTiO_3$. It is evident that the $BaTiO_3$ fiber is embedded within the PVDF matrix and aligned along its fiber axis. The inset in the figure shows a higher magnification image taken of another fiber area.

crystals in PVDF matrix. However, the content of α- and β-crystalline structures inside the PVDF fibers due to the $BaTiO_3$ inclusion is estimated using the Beer–Lambert law. Thus, according to Beer-Lambert law, α- and β-crystalline content can be estimated by using the absorbencies for α- and β-crystals at 766 and 840 cm^{-1}, respectively. The fraction of β-phase is calculated from [25, 26]:

$$F(\beta) = \frac{A_\beta}{(1.26\,A_\alpha + A_\beta)} \tag{1}$$

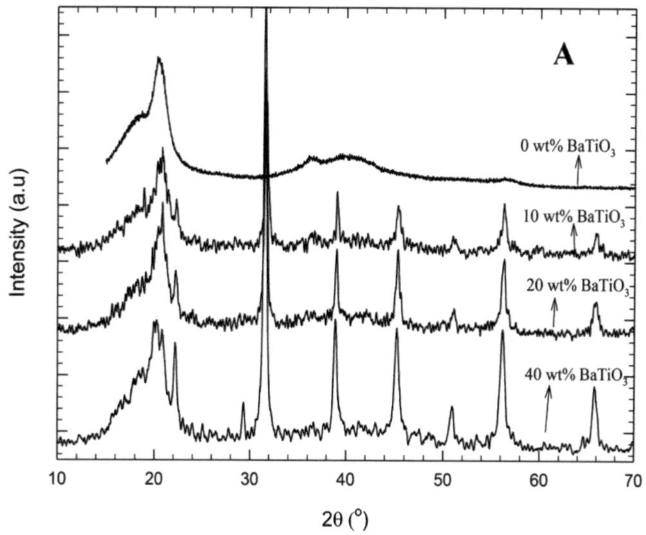

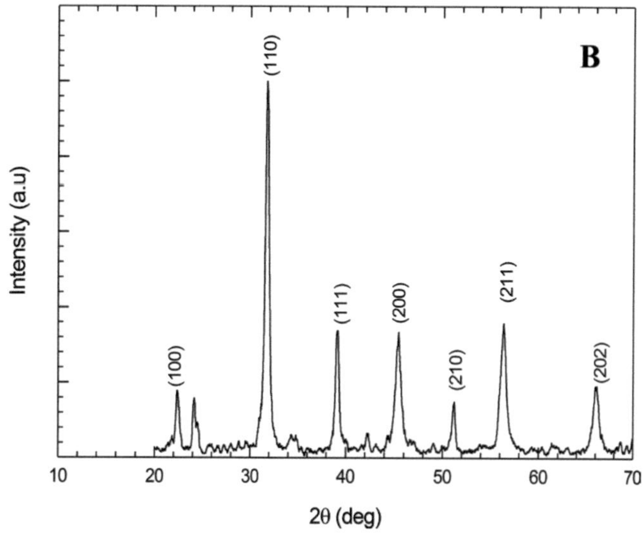

Figure 3. (A) XRD patterns of PVDF fibers as a function of $BaTiO_3$ content. All peaks corresponding to the crystalline phases of PVDF and $BaTiO_3$ are evident in the fibers reinforced with $BaTiO_3$. (B) XRD pattern of neat $BaTiO_3$ to identify the cubic and tetragonal structure of $BaTiO_3$.

where A_α and A_β are the corresponding absorbency at 766 and 840 cm^{-1}. **Figure 4B** shows the fraction of β-crystals estimated within the PVDF matrix for Sample 2 fibers. The β-crystals content within the sample fibers increases from 81% estimated for neat fibers to 87% for fibers reinforced with 40 wt% $BaTiO_3$. This shows that the inclusion of $BaTiO_3$ improves the ferroelectric crystal phase formation within the PVDF fibers.

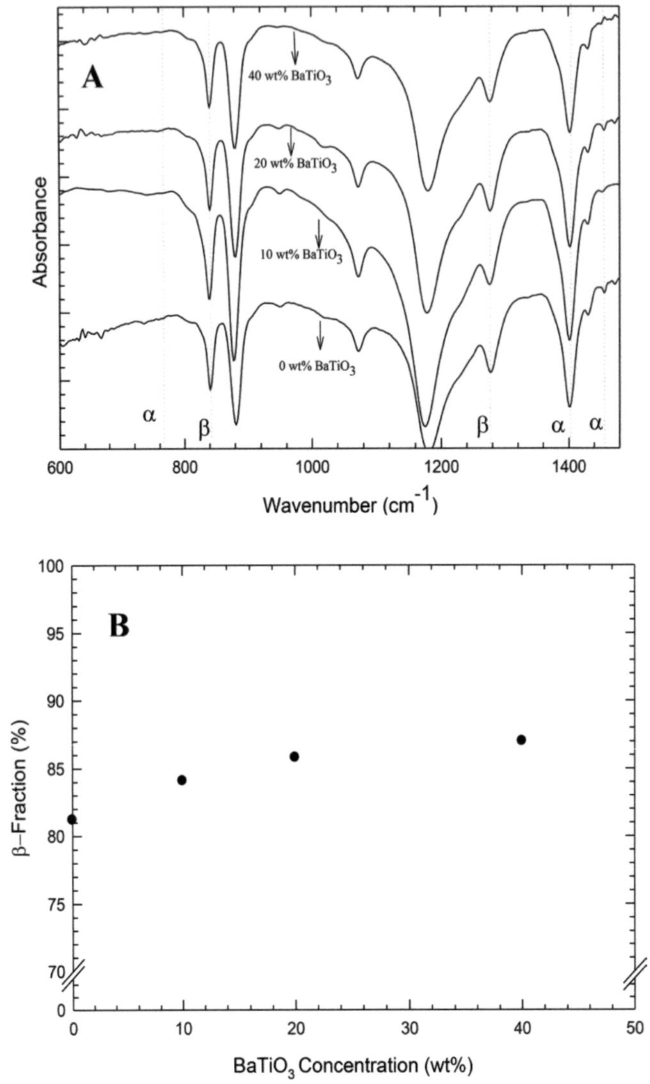

Figure 4. (A) IR spectra of fibers as a function of $BaTiO_3$ content. α- and β-related bands of PVDF are indexed in the spectra. (B) Fraction of β-phase as a function of $BaTiO_3$ content. The β-phase fractions are determined using Eq. (1).

The FTIR data corroborates the XRD results. The improvement in the fraction of β-crystals within PVDF can be attributed to the changes in crystalline structure development due to the inclusion of $BaTiO_3$. It proves that the $BaTiO_3$ content in PVDF fibers plays an important role in influencing the crystallization of PVDF and also promotes phase change within the PVDF. These results on crystallization induced by $BaTiO_3$ content are in agreement with the results reported by Dang et al. [17]. They show that $BaTiO_3$ can be easily absorbed on surfaces of PVDF due to the presence of interstitial hydrogen ion $BaTiO_3$ lattice. This also helps in the dispersion of $BaTiO_3$ in polar DMF solvent. The homogeneous dispersion and absorption of $BaTiO_3$ on the surface of PVDF plays an important role to induce crystal structure changes [17].

We will now characterize the deformation behavior of these PVDF fibers filled with $BaTiO_3$ and also illustrate the mechanisms behind the improved mechanical integrity of the composite fibers. **Figure 5A** shows typical stress–strain curves for $BaTiO_3$/PVDF fibers. The break strain decreases with increasing $BaTiO_3$ content within the fiber. **Figure 5B** shows the tensile modulus and tensile strength of the composite fibers. In general, as the $BaTiO_3$ content increases, the composite fibers become stiffer and stronger. The stiffness and strength of neat PVDF fibers are increased by 36 and 12%, respectively, when they are filled with 10 wt% $BaTiO_3$ content. These tensile results of $BaTiO_3$/PVDF fibers are contrary to those obtained by Fang et al. [16] on thin films of $BaTiO_3$/poly(vinyledene fluoride-trifluoroethylene) (P(VDF-TrFE)) [16]. In their study, adding $BaTiO_3$ to P(VDF-TrFE) softened and reduced the matrix tensile strength. Reductions in both stiffness

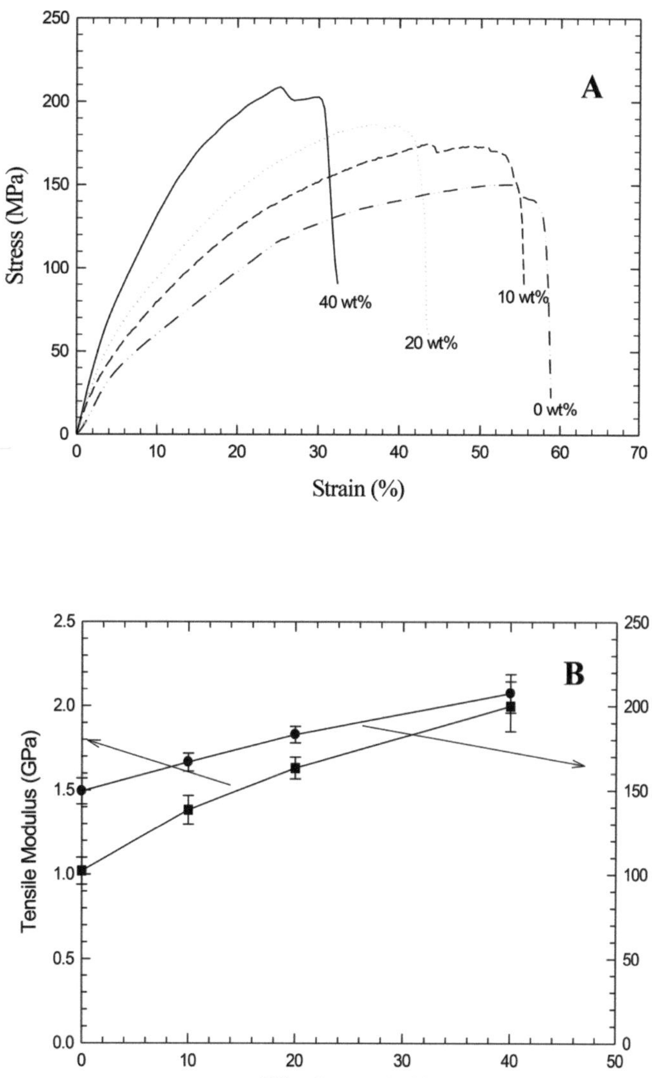

Figure 5. (A) Stress-strain curves obtained from tensile tests conducted on fiber samples as a function of $BaTiO_3$ concentration. (B) Plot of tensile strength and modulus of fiber samples as a function of $BaTiO_3$ content.

and strength of the composite films are attributed to the weak chemical bonding between the $BaTiO_3$ particles and P(VDF-TrFE) matrix and the inhibition of crystallinity due to the addition of $BaTiO_3$ particles. By contrast, we attribute the increase in tensile properties of $BaTiO_3$/PVDF fibers to the reinforcement effect of $BaTiO_3$. We employ DMA to investigate the influence of $BaTiO_3$ on the polymeric chain mobility within the fibrous matrix and attribute the increase in strength and stiffness of the fibers to the reinforcement effect of $BaTiO_3$. **Figure 6** presents the loss tangent (tan δ) as a function of temperature for $BaTiO_3$/PVDF fibers. The β-transition region of the fibers relates to the glass transition temperature (T_g) of the fibers. Comparing the effect of $BaTiO_3$ content on the fibrous matrix, neat PVDF fibers show a peak of the tan δ *versus* temperature curve at the lowest temperature (−37.33°C) while the composite fibers show peaks of the tan δ *versus* temperature curves at higher temperatures. For example, fibers filled with 10, 20 and 40 wt% $BaTiO_3$ show T_g at −36.06, −35.44 and −31.9°C, respectively. This explains that the mobility of neighboring chains surrounding the $BaTiO_3$ phase is inhibited. In neat PVDF fibers, the chains are relatively free to rotate while in fibers filled with $BaTiO_3$ phase, the $BaTiO_3$ freezes the movement of the chains. The fact that $BaTiO_3$ are adsorbed on PVDF surface may also explain the $BaTiO_3$ inclusion restricts the mobility of PVDF chains.

The effect of $BaTiO_3$ content on the relative permittivity of Sample 2 fibers has been studied over a broad frequency range from the measured capacitance (C) which is given by:

$$\varepsilon_r = \frac{Cd}{A\varepsilon_0} \tag{2}$$

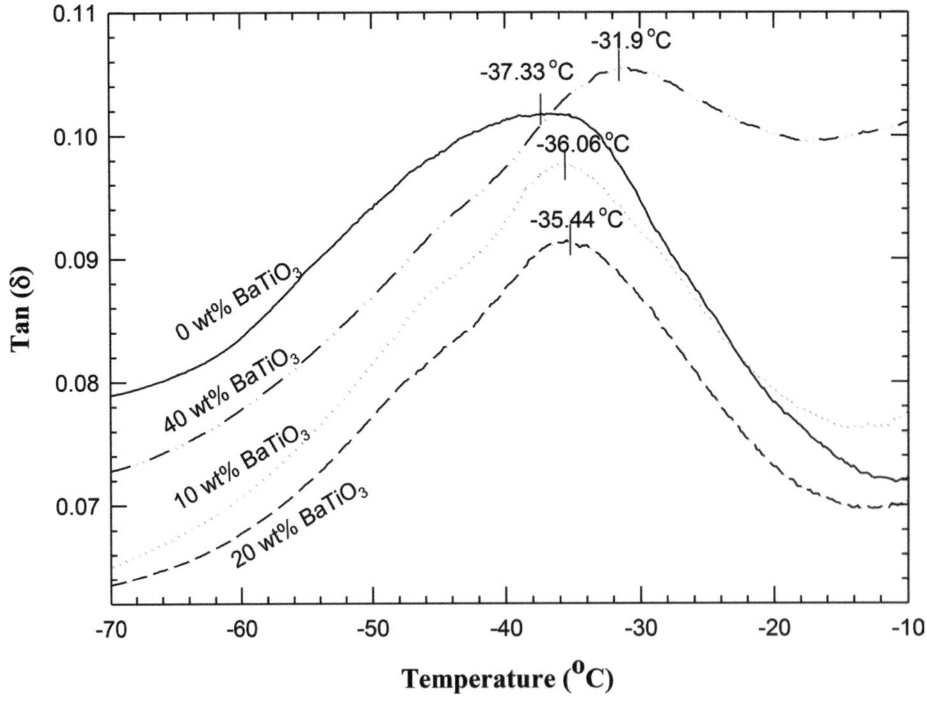

Figure 6. Plot of tan δ versus temperature obtained using DMA. The effect of $BaTiO_3$ content on the glass transition temperature (T_g) of the fiber is evaluated.

where ε_r is relative permittivity of capacitor, d thickness of samples, A surface area, and ε_0 is free space dielectric constant taken as 8.854×10^{-12} F/m. The dependence of relative permittivity, ε_r, and dielectric loss tangent of $BaTiO_3$/PVDF fibers on frequency are shown in **Figure 7A** and **B**, respectively. Clearly, the dielectric permittivity increases with $BaTiO_3$ content; but the dielectric

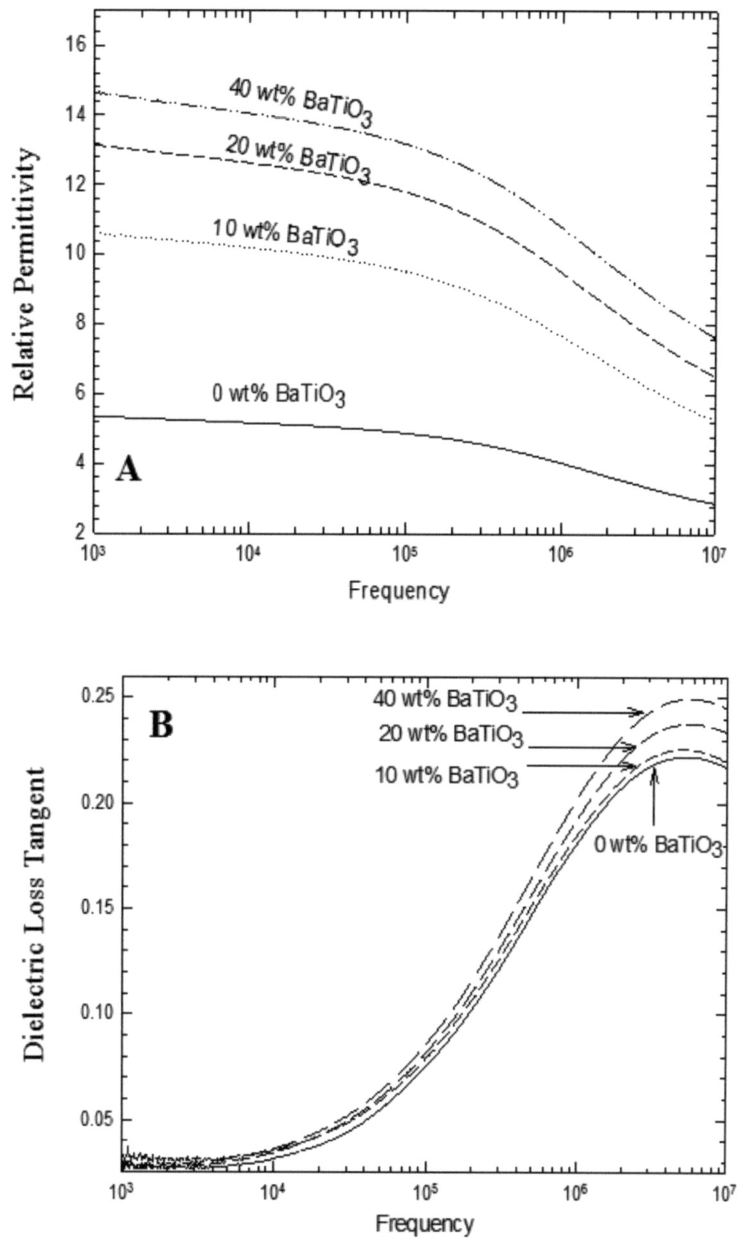

Figure 7. (A) Frequency dependence of effective dielectric constant (measured at 300 K) of PVDF fiber as function of $BaTiO_3$ content. (B) Frequency dependence of effective dielectric loss tangent (measured at 300 K) of PVDF fiber as functions of $BaTiO_3$ content.

loss tangent remains nearly the same for all fibers. The relaxation drop in relative permittivity at ~10^5 Hz is due to the characteristic dielectric behavior of PVDF matrix, and the rapid drop in dielectric permittivity after 10^5 Hz is because the dipole relaxations of the fibers lag behind the fast change of the applied field [1, 4, 17]. Finally, the relative permittivity of fibers with 0, 10, 20 and 40 wt% $BaTiO_3$ at 10^3 Hz are 5.4, 10.6, 13.13 and 14.6, respectively (see **Figure 7A**).

Typically, the dielectric loss of a material should be as low as possible for its use in capacitor applications. **Figure 7B** compares the dielectric loss tangent of the sample fibers as a function of frequency. Typically, the loss tangent of $BaTiO_3$ reaches a maximum value in the gigahertz and terahertz frequency range and it does not show any significant dielectric losses up to the megahertz frequency range. Thus, the dielectric loss of PVDF fibers filled with $BaTiO_3$ recorded at low frequency (up to 10^6 Hz) is mainly attributed to the loss tangent values of PVDF. A moderate increase in the loss tangent above 10^6 Hz frequency is due to the loss tangent contribution arising from the $BaTiO_3$. It is evident from **Figure 7B** there is a clear dielectric loss peak at 5 MHz which is attributed to the relaxation loss process of PVDF [17]. These results show that the dielectric properties of $BaTiO_3$/PVDF are useful for applications in electronic devices.

4. Conclusion

In this study, we use electrospinning to obtain PVDF fibers reinforced with $BaTiO_3$. We demonstrate the effect of $BaTiO_3$ on the crystalline structure developments, tensile and dielectric properties of PVDF fibers. Reinforcing PVDF fibers with $BaTiO_3$ promotes the formation of ferroelectric β-crystalline phase within the fibers. Tensile strength and stiffness of the fibers increase with $BaTiO_3$ content. Finally, the effective dielectric permittivity and dielectric loss of the fibers increase with $BaTiO_3$ content at all frequencies studied.

Acknowledgements

We thank the Australian Research Council (ARC) for the continuous support of this project (#DP0665856).

Author details

Avinash Baji[1]* and Yiu-Wing Mai[2]

*Address all correspondence to: avinash_baji@sutd.edu.sg

1 Engineering Product Development (EPD) Pillar, Singapore University of Technology and Design (SUTD), Singapore

2 Centre for Advanced Materials Technology (CAMT), School of Aerospace, Mechanical and Mechatronic Engineering, The University of Sydney, Sydney, NSW, Australia

References

[1] Bai Y, Cheng ZY, Bharti V, Xu HS, Zhang QM. High-dielectric-constant ceramic-powder polymer composites. Applied Physics Letters. 2000;**76**:3804-3806

[2] Baji A, Mai Y-W, Li Q, Liu Y. Nanoscale investigation of ferroelectric properties in electrospun barium titanate/polyvinylidene fluoride composite fibers using piezoresponse force microscopy. Composites Science and Technology. 2011;**71**:1435-1440

[3] Li JJ, Seok SI, Chu BJ, Dogan F, Zhang QM, Wang Q. Nanocomposites of ferroelectric polymers with TiO(2) nanoparticles exhibiting significantly enhanced electrical energy density. Advanced Materials. 2009;**21**:217

[4] Damjanovic D. Ferroelectric, dielectric and piezoelectric properties of ferroelectric thin films and ceramics. Reports on Progress in Physics. 1998;**61**:1267-1324

[5] Takenaka T, Nagata H. Current status and prospects of lead-free piezoelectric ceramics. Journal of the European Ceramic Society. 2005;**25**:2693-2700

[6] Baji A, Mai Y-W, Li QA, Wong SC, Liu Y, Yao QW. One-dimensional multiferroic bismuth ferrite fibers obtained by electrospinning techniques. Nanotechnology. 2011;**22**:235702

[7] Buckley J, Cebe P, Cherdack D, Crawford J, Ince BS, Jenkins M, et al. Nanocomposites of poly(vinylidene fluoride) with organically modified silicate. Polymer. 2006;**47**:2411-2422

[8] Gregorio R, Cestari M. Effect of crystallization temperature on the crystalline phase content and morphology of poly(vinylidene fluoride). Journal of Polymer Science Part B: Polymer Physics. 1994;**32**:859-870

[9] Kim P, Doss NM, Tillotson JP, Hotchkiss PJ, Pan MJ, Marder SR, et al. High energy density nanocomposites based on surface-modified BaTiO(3) and a ferroelectric polymer. ACS Nano. 2009;**3**:2581-2592

[10] Tang HX, Lin YR, Andrews C, Sodano HA. Nanocomposites with increased energy density through high aspect ratio PZT nanowires. Nanotechnology. 2011;**22**:015702

[11] Andrews C, Lin Y, Sodano HA. The effect of particle aspect ratio on the electroelastic properties of piezoelectric nanocomposites. Smart Materials and Structures. 2010;**19**:025018

[12] Newnham RE, Skinner DP, Cross LE. Connectivity and piezoelectric-pyroelectric composites. Materials Research Bulletin. 1978;**13**:525-536

[13] Dias CJ, DasGupta DK. Inorganic ceramic/polymer ferroelectric composite electrets. IEEE Transactions on Dielectrics and Electrical Insulation. 1996;**3**:706-734

[14] Chen LF, Hong YP, Chen XJ, Wu QL, Huang QJ, Luo XT. Preparation and properties of polymer matrix piezoelectric composites containing aligned $BaTiO_3$ whiskers. Journal of Materials Science. 2004;**39**:2997-3001

[15] Shen Y, Nan CW, Li M. Anisotropic electrical properties of semiconductive bi(2)S(3) nanorod filled ferroelectric polyvinylidene fluoride. Chemical Physics Letters. 2004;**396**: 420-423

[16] Fang F, Yang W, Zhang MZ, Wang Z. Mechanical response of barium-titanate/polymer 0-3 ferroelectric nano-composite film under uniaxial tension. Composites Science and Technology. 2009;**69**:602-605

[17] Dang ZM, Wang HY, Zhang YH, Qi JQ. Morphology and dielectric property of homogenous $BaTiO_3$/PVDF nanocomposites prepared via the natural adsorption action of nanosized $BaTiO_3$. Macromolecular Rapid Communications. 2005;**26**:1185-1189

[18] Yuh J, Nino JC, Sigmund WM. Synthesis of barium titanate ($BaTiO_3$) nanofibers via electrospinning. Materials Letters. 2005;**59**:3645-3647

[19] Kamalasanan MN, Kumar ND, Chandra S. Structural and microstructural evolution of barium titanate thin-films deposited by the sol-gel process. Journal of Applied Physics. 1994;**76**:4603-4609

[20] Frey MH, Payne DA. Grain-size effect on structure and phase transformations for barium titanate. Physical Review B. 1996;**54**:3158-3168

[21] Lu SW, Lee BI, Wang ZL, Samuels WD. Hydrothermal synthesis and structural characterization of $BaTiO_3$ nanocrystals. Journal of Crystal Growth. 2000;**219**:269-276.22

[22] Zheng JF, He AH, Li JX, Han CC. Polymorphism control of poly(vinylidene fluoride) through electrospinning. Macromolecular Rapid Communications. 2007;**28**:2159-2162

[23] Priya L, Jog JP. Poly(vinylidene fluoride)/clay nanocomposites prepared by melt intercalation: Crystallization and dynamic mechanical behavior studies. Journal of Polymer Science Part B: Polymer Physics. 2002;**40**:1682-1689

[24] Levi N, Czerw R, Xing SY, Iyer P, Carroll DL. Properties of polyvinylidene difluoride-carbon nanotube blends. Nano Letters. 2004;**4**:1267-1271

[25] Gregorio R, Cestari M. Effect of crystallization temperature on the crystalline phase content and morphology of poly(vinylidene fluoride). Journal of Polymer Science Part B: Polymer Physics. 1994;**32**:859-870

[26] Buckley J, Cebe P, Cherdack D, Crawford J, Ince BS, Jenkins M, et al. Nanocomposites of poly(vinylidene fluoride) with organically modified silicate. Polymer. 2006;**47**:2411-2422